Lecture Notes in Economics and Mathematical Systems

446

Springer
Berlin
Heidelberg
New York
Barcelona
Budapest
Hong Kong
London
Milan
Paris
Santa Clara
Singapore
Tokyo

Geert Dhaene

Encompassing

Formulation, Properties and Testing

 Springer

Author

Dr. Geert Dhaene
Katholieke Universiteit Leuven
Centrum voor Economische Studiën
Naamsestraat 69
B-3000 Leuven, Belgium

Library of Congress Cataloging-in-Publication Data

Dhaene, Geert, 1959-
    Encompassing : formulation, properties and testing / Geert Dhaene.
        p.   cm. -- (Lecture notes in economics and mathematical
    systems; 446)
    Includes bibliographical references (p.   ) and indexes.
    ISBN 3-540-61890-2 (Springer-Verlag Berlin Heidelberg New York:
acid-free paper)
    1. Statistical hypothesis testing. 2. Social sciences-
-Statistical methods.   3. Social sciences--Mathematical models.
I. Title. II. Series.
HA31.5.D47   1997
300'.1'5195--dc21
                                                              96-36816
                                                                 CIP

ISSN 0075-8442
ISBN 3-540-61890-2 Springer-Verlag Berlin Heidelberg New York

© Springer-Verlag Berlin Heidelberg 1997
Printed in Germany

Typesetting: Camera ready by author
SPIN: 10546684   42/3142-543210 - Printed on acid-free paper

*Altogether then, I said, such men
would believe the truth to be
nothing else than the shadows
of the artifacts?*
PLATO, *The Republic,* VII, 515.

Acknowledgements

Through their intellectual influence on me, many people have contributed to the writing of this monograph. I cannot mention all their names, but some stand out.

Anton Barten, who taught me econometrics, has undoubtedly been the most influential. I recall as a Ph.D. student being fascinated by his high standards of lucidity and mathematical elegance. Without him, I probably would never have been attracted to the subject.

Christian Gourieroux, at a later stage, encouraged me to continue this research. Apart from his numerous invaluable comments, the influence of his work with Alain Monfort will be obvious throughout this monograph.

Erik Schokkaert had a pervasive impact on me, both humanly and intellectually. He made me aware that achieving elegance and relevance *simultaneously* can be unduly hard.

Over the years I have immensely benefited from discussions with Denis de Crombrugghe. It is only fair to say that, without him, this monograph may simply not have been written.

I am also grateful to Jan Beirlant, Guido De Bruyne, Casper de Vries, and Frans Spinnewijn, who, in addition to the people already mentioned, have commented upon earlier drafts.

The writing of this monograph started at the Katholieke Universiteit Leuven, and continued while I was Human Capital and Mobility Fellow at the Tinbergen Institute, Rotterdam, and Post-Doctoral Researcher at the National Fund for Scientific Research, Belgium. I also spent some months at CREST, Paris. Financial support from these institutions is gratefully acknowledged. I thank Dr. Werner Müller, Executive Economics Editor at Springer-Verlag, for his repeated patience—to such a point that it made me feel truely ashamed.

A final word of thanks goes to my parents for their support, and, above all, to my sons Klaas and Bavo, my *sine qua non*.

Geert Dhaene

Contents

Introduction

The history of many sciences is characterized by an almost continuous emergence of new theories. From a normative point of view, the survival of a new theory should mainly be determined by its ability to explain a new body of facts which the existing theories are unable to explain. If in addition the new theory is able to explain all the results obtained by the existing theories and if it can point out why these theories fail to explain certain facts, it should become the dominant theory. Otherwise, it might coexist with other theories for some time. Hence, a new theory ought to be judged not only by confronting it with existing facts, but also by confronting it with existing theories.

The idea that a theory should be able to account for the results obtained by other theories, although implicitly adhered to by many scientists, has rarely been formalized. The statistics literature on parametric hypothesis testing, though, might be seen as an instance of such a formalization. Theories are then formulated by means of statistical models, and there is a standard methodology to test more specific models against more general ones. This is, however, unsatisfactory for at least two reasons. First, with a few notable exceptions, the classical theory of parametric inference relies on the assumption that the more general model is correctly specified. In most, if not all, practical situations of models with non-experimental data, this assumption is all too strong. Secondly, it rarely occurs in practice that one of two competing theories is more general than the other, in the sense of reducing to the other in special cases. Hence, there is a general need for procedures to test theories where neither one can be reduced to the other. Such procedures, sometimes referred to as non-nested hypotheses tests, have been developed from the early sixties onwards, but they remain subject to the first criticism. In particular, they rely on the assumption that at least one of the models is correctly specified.

The encompassing principle (see, e.g., Hendry–Richard [1982, 1983, 1990], Mizon [1984], Mizon–Richard [1986] and Gouriéroux–Monfort [1995]), which was introduced into the econometrics literature, offers a more formal basis for analyzing whether a theory is able to account for the results obtained by other theories. As in classical statistical inference, theories are

1

phrased in terms of statistical models. The fundamental idea of the encompassing principle is that, if we knew the underlying mechanism generating the real world data, then we would be able to predict correctly (in statistical terms) the results of any model, no matter how good or bad this model is. Hence, the finding that a model, which by definition is designed to mimic some part of this mechanism, is able to predict the results of another model correctly, is evidence in favour of the former model, since at least it can mimic this fundamental property of the underlying mechanism. The former model is then said to encompass the latter one. Conversely, if a model fails to predict the results of another model correctly, it reveals shortcomings in mimicking the underlying mechanism or, for that sake, reality, in which case the former model does not encompass the latter one. Furthermore, two models may at the same time encompass each other, or at the same time fail to encompass the other one. Note that this line of thought does not assume that any of the models is correctly specified, nor that the models can be ordered according to their generality. These features are characteristic of the encompassing approach.

This monograph seeks to contribute to the theory of encompassing. In general, it describes a framework for a formal comparison of a pair of empirical parametric models and develops a methodology for hypothesis testing. Throughout, the pair of models is arbitrary in the sense that the models may be nested, disjoint, or overlapping, and that any of the models may or may not be correctly specified. However, attention is limited to models and data generating processes with independent and identically distributed observations.

Chapter 1 is entirely devoted to the notion of pseudo-true values, which is a basic concept in the encompassing framework. The pseudo-true value of a model is the parameter value which identifies the distribution from the model that is closest to the true data generating process according to the Kullback-Leibler information criterion. The interest in this notion is motivated by the well known fact that the maximum likelihood estimator of the parameter converges almost surely to the pseudo-true value. We give regularity conditions ensuring its existence and uniqueness, discuss its determination in the presence of location and scale parameters, and provide several detailed examples for unconditional and conditional models.

2

In Chapter 2, we study the encompassing relation. It is defined as a binary relation on a given class of parametric models. Building on the work of Gouriéroux–Monfort [1995], we investigate the properties of this relation and discuss its usefulness for empirical model building. Our main conclusions are that the hypothesis that one model encompasses another model is weaker than the hypothesis that the former model is correctly specified, that the encompassing relation is not transitive, and that the encompassing relation, combined with the parsimony principle, can be used as a meaningful model reduction device. We also reconsider the examples given in Chapter 1 and study the encompassing relation in the context of these examples.

The problem of statistical inference with respect to the encompassing relation is addressed in Chapter 3, which is the longest chapter and contains most of the new material. The interest in this problem stems from the critical role that can be attributed to the encompassing principle in a modelling strategy and from the fact that the encompassing hypothesis can be viewed as a robust hypothesis (i.e., not subject to a maintained hypothesis), being applicable to nested, disjoint, or overlapping models. As such, testing for encompassing unifies the literature on robust nested hypotheses testing and the literature on non-robust non-nested hypotheses testing. The distribution theory underlying the encompassing tests is to be developed on the basis of the unknown true data generating process. This difficulty seems to have been overlooked in the early literature on encompassing. It can be overcome by noting that the relevant aspects of this process are gradually revealed by the data generated from it, and can therefore be consistently estimated. Gouriéroux–Monfort [1992], later published as Gouriéroux–Monfort [1995], and Smith [1993] and White [1994] have developed a methodology for testing the encompassing hypothesis without assuming that the true data generating process lies within one of the specified models. In a general dynamic context, these authors derived the limit distributions of appropriate encompassing test statistics under the assumption that the encompassing hypothesis is true. We extend their distributional results to the case where the encompassing hypothesis does not hold, though in the less general static context.

Before the development of the distribution theory, we put the encompassing hypothesis and the testing methodology into perspective with other statistical hypotheses and other approaches to parametric inference. Furthermore we show that, under certain regularity conditions, the encompassing hypothesis can be restated in terms of almost sure limits of some important statistics of the data, all of which relate to the three main principles in classical parametric inference: the Wald principle, the score or Lagrange multiplier principle, and the likelihood ratio principle. We define a modified likelihood ratio which is shown to be more appropriate for testing the encompassing hypothesis than the usual one. Under additional regularity assumptions, and as a prerequisite to the construction of the tests, we characterize the limit distributions of the Wald and score vectors, of the modified likelihood ratio, and of the relevant quadratic forms in the Wald and score vectors. It turns out that the form of the limit distributions of these statistics depends critically on whether the encompassing hypothesis holds or not. Specifically, the limit distributions are weighted sum of chi-squares distributions if the encompassing hypothesis holds, and normal distributions otherwise. Furthermore, the statistics are bounded in probability if the encompassing hypothesis holds, and converge almost surely to infinity otherwise. Although we shall not be concerned with power analysis, we note that these results provide a basis for a power analysis under fixed alternatives. We deduce consistent encompassing tests with correct asymptotic size from these results. The special cases of nested models and hypotheses in implicit form allow simpler results and are therefore presented separately. By necessity, all the results are of an asymptotic nature.

Chapter 4 gives an extensive discussion of the encompassing relation on the class of normal linear models, both from a conceptual and an inferential point of view. In this class of models, encompassing is characterized by an orthogonality condition between a regression error and the regressors of the other model. Moreover, the notion of a comprehensive model, which is particularly appealing in the context of non-nested regression models and has inspired several methods of inference, is shown to be closely related to the principle of encompassing. In fact we prove that, in the class of normal linear models, encompassing a model is equivalent to encompassing the comprehensive model. The general distribution theory developed in Chapter 3 is applied to yield robust Wald, score and likelihood ratio encompassing tests for nested or non-nested normal linear models. The gen-

4

eral conclusions made there do not specialize further: appropriate quadratic forms in the Wald and score vectors are, under the encompassing hypothesis, asymptotically equivalent and distributed as a chi-square variate, whereas the modified likelihood ratio is asymptotically distributed as a weighted sum of chi-squares. Whether or not the models are nested or non-nested is immaterial to the form of the limiting distributions obtained. The proposed tests are invariant under non-singular linear transformations of the models involved.

We conclude with a summary of the main results.

Chapter 1

Pseudo-true values

1. Introduction

This preliminary chapter discusses the issue of approximating the distributions of a parametric family $\mathcal{F} = \{F_\alpha \,|\, \alpha \in \Omega_\mathcal{F} \subset R^m\}$ by those of another parametric family $\mathcal{G} = \{G_\beta \,|\, \beta \in \Omega_\mathcal{G} \subset R^n\}$. The definition of encompassing will be based on such approximations, which are therefore discussed at some length here. The general problem consists of finding the mapping of $\mathcal{F}$ into $\mathcal{G}$ which associates with each distribution $F_\alpha \in \mathcal{F}$ the distribution $G_\beta \in \mathcal{G}$ closest to it according to some criterion. The adoption of the Kullback-Leibler [1951] Information Criterion (KLIC) as a distance measure between distributions defines such a mapping. The mapping which results from this choice will be the sole object of study in this chapter. In line with Sawa [1978], we call the values obtained under this mapping pseudo-true distributions, and their associated parameter vectors pseudo-true (parameter) values. Although only implicitly, pseudo-true values seem to have appeared first in the pioneering work of Cox [1961, 1962] in connection with non-nested hypothesis testing.

In Section 2, the KLIC is defined and some of its properties are reviewed. Following Rényi [1961], the choice of the KLIC as a criterion function may be justified on axiomatic grounds. The pseudo-true value of a parameter vector is then defined as the solution of a minimization problem involving the KLIC. We also touch briefly upon the role of the pseudo-true value in the theory of pseudo-maximum likelihood (pseudo-ML) estimation, i.e., the theory of ML estimation of possibly misspecified models. A fuller account of this theory is given in Chapter 3.

Section 3 discusses two ways of extending the approach to families of conditional distributions so as to allow for explanatory variables. One may either redefine the pseudo-true value with reference to a marginal distribution of conditioning variables, or one may consider the conditioning variables as (a sequence of) fixed numbers, and redefine the pseudo-true value appropriately.

In Section 4, we discuss the determination of the pseudo-true value in the presence of location or scale parameters.

We give several examples in Section 5 and some concluding remarks in Section 6.

Throughout, we have made an attempt to provide appropriate regularity conditions. They are stated using standard measure theoretic concepts as discussed, for example, in Halmos [1974] or Kingman–Taylor [1977]. The regularity conditions given in this chapter are sufficient to guarantee the existence and uniqueness of the pseudo-true value. Most authors present such conditions as part of a stronger set of assumptions ensuring also existence, almost sure convergence to the pseudo-true value and asymptotic normality of the pseudo-ML estimator—see, e.g., Huber [1967], White [1982] and Vuong [1989]. Such assumptions will be presented separately in Chapter 3.

2. Pseudo-true values

We consider a σ-finite measure space (A, σ_A, ν) where A is the Euclidean space R^k, σ_A is the Borel σ-field on A and ν is a σ-finite measure on (A, σ_A). The measure ν and the families of distributions $\mathcal{F} = \{F_\alpha | \alpha \in \Omega_\mathcal{F} \subset R^m\}$ and $\mathcal{G} = \{G_\beta | \beta \in \Omega_\mathcal{G} \subset R^n\}$, also defined on (A, σ_A), satisfy the following assumptions. (Occasionally, we use the word 'parametrization' to refer to the way in which the elements of a parametric family, say $\mathcal{F}$, are indexed. Formally, a parametrization is a bijective map of $\mathcal{F}$ onto a subset of a Euclidean space. This subset is called the parameter space. A reparametrization is a bijective map of the parameter space onto a subset of a Euclidean space, which may or may not coincide with the former parameter space.)

Assumption A1. (a) For every $\alpha \in \Omega_\mathcal{F}$, F_α has a density f_α relative to ν. (b) f_α is continuous in α.

Assumption A2. (a) For every $\beta \in \Omega_\mathcal{G}$, G_β has a density g_β relative to ν. (b) g_β is continuous in β. (c) For every $\beta \in \Omega_\mathcal{G}$, ν is absolutely continuous with respect to G_β.

Under Assumption A1-(a), F_α is absolutely continuous with respect to ν, i.e., sets of ν-measure zero are also sets of F_α-measure zero. Assumptions

8

A2-(a) and A2-(c) taken together imply that ν and G_β are absolutely continuous with respect to each other, i.e., the sets of ν-measure zero coincide with the sets of G_β-measure zero. As a result of these assumptions, the support of g_β does not depend on β, and it always contains the support of f_α. Note furthermore that, since ν need not be the Lebesgue measure, F_α and G_β may be continuous, discrete, or mixed.

Consider now a random vector Y taking its values in A and whose distribution is F_α. Given another distribution G_β, one may ask what is the distance between F_α and G_β. A directional measure of this distance is provided by the Kullback-Leibler [1951] Information Criterion (KLIC), which is defined as

$$(2.1) \qquad I(F_\alpha, G_\beta) = E_\alpha \log \frac{f(Y; \alpha)}{g(Y; \beta)},$$

where $f(\cdot; \alpha) = f_\alpha(\cdot)$, $g(\cdot; \beta) = g_\beta(\cdot)$ and E_α denotes the mathematical expectation relative to F_α. It is worth pointing out that the existence of $I(F_\alpha, G_\beta)$ requires the support of f_α to be contained in the support of g_β, whence the need of Assumption A2-(c). From Jensen's inequality, it follows that

$$(2.2) \qquad E_\alpha \log \frac{f(Y; \alpha)}{g(Y; \beta)} \geq -\log E_\alpha \frac{g(Y; \beta)}{f(Y; \alpha)} \geq 0,$$

showing that the KLIC is non-negative and is zero if and only if $F_\alpha = G_\beta$. Although the KLIC is not a metric since the triangle inequality does not hold and in general $I(F_\alpha, G_\beta) \neq I(G_\beta, F_\alpha)$, we shall call $I(F_\alpha, G_\beta)$ the distance between F_α and G_β, bearing in mind the order in which the distributions are mentioned.

The use of the KLIC as a distance function between distributions may be justified by a number of results from information theory. Upon its introduction, Kullback–Leibler [1951] presented the KLIC as a generalization of Shannon's [1948] and Wiener's [1948] definition of entropy of a distribution, viz.

$$(2.3) \qquad H(G_\beta) = -E_\beta \log g(Y; \beta),$$

where E_β denotes the mathematical expectation relative to G_β. The entropy $H(G_\beta)$ is a measure of the uncertainty concerning the outcome of an

9

experiment which has distribution G_β. As a measure of uncertainty, entropy is conceptually dual to the information which is revealed by observing the outcome of the experiment. Therefore, $H(G_\beta)$ is also a measure of (average) information of G_β. A fundamental property of entropy is its additivity: the entropy of a combined experiment which consists of two independent experiments is equal to the sum of the entropies of the individual experiments. The requirement of additivity plays a key role in axiomatic approaches in information theory—see, e.g., Shannon [1948], Wiener [1948], Barnard [1951] and Fadeev [1957] for axiomatizations of the entropy of discrete distributions. Turning back to the KLIC, Rényi [1961] has given a similar set of axioms which characterize the KLIC as a measure of the information concerning Y which is revealed by observing some event $E \in \sigma_A$, such that G_β is the unconditional distribution of Y and F_α is the conditional distribution of Y, given that E has occurred. In other words, the KLIC may be interpreted as a measure of the information gain associated with the transition from G_β to the more informative F_α. Notice that in this setting F_α is absolutely continuous with respect to G_β, as was assumed. The purpose is now to associate with each distribution $F_\alpha \in \mathcal{F}$ the distribution $G_\beta \in \mathcal{G}$ for which the KLIC is minimal. Intuitively, this procedure minimizes the uncertainty with respect to Y if one does not dispose of the observation of the event E, as it minimizes the information gain that would result from observing it. These observations give some justification for the KLIC as a distance function between distributions.

Although more restrictive than strictly necessary, we make an assumption to ensure that the KLIC is finite for every pair (F_α, G_β).

Assumption A3. *For every $\alpha \in \Omega_\mathcal{F}$, $E_\alpha \log f(Y; \alpha)$ exists and $|\log g(y; \beta)|$ is dominated by an F_α-integrable function independent of β.*

For every $\alpha \in \Omega_\mathcal{F}$, the KLIC, viewed as a function of β, is real-valued and continuous on $\Omega_\mathcal{G}$ under Assumptions A1–A3. We impose the existence and uniqueness of a minimum of the KLIC with respect to β.

Assumption A4. *For every $\alpha \in \Omega_\mathcal{F}$, $I(F_\alpha, G_\beta)$ has a unique minimum with respect to $\beta \in \Omega_\mathcal{G}$.*

Given Assumption A3, it follows from (2.1) and (2.3) that

$$(2.4) \qquad I(F_\alpha, G_\beta) = -H(F_\alpha) - E_\alpha \log g(Y; \beta),$$

10

and minimizing the KLIC with respect to β is seen to be equivalent to maximizing $E_\alpha \log g(Y; \beta)$ with respect to β. By Assumption A4, this problem has a unique solution. Consequently, the mapping

$$(2.5) \qquad\qquad \tau_{\alpha\beta} : \Omega_{\mathcal{F}} \to \Omega_{\mathcal{G}} : \alpha \mapsto \beta_\alpha,$$

where β_α is the unique solution of

$$(2.6) \qquad\qquad \max_{\beta \in \Omega_{\mathcal{G}}} E_\alpha \log g(Y; \beta),$$

associates with each value $\alpha \in \Omega_{\mathcal{F}}$ a value $\beta_\alpha \in \Omega_{\mathcal{G}}$ such that the distance between F_α and G_{β_α} is minimal according to the KLIC. Therefore, β_α is called the *pseudo-true value* of β relative to F_α (see, e.g., Sawa [1978] and White [1982]). The induced mapping of $\mathcal{F}$ into $\mathcal{G}$ is

$$(2.7) \qquad\qquad \tau_{FG} : \mathcal{F} \to \mathcal{G} : F_\alpha \mapsto G_{\beta_\alpha},$$

where, in the same spirit, we call G_{β_α} the *pseudo-true distribution* of $\mathcal{G}$ relative to F_α.

Based on the KLIC, the distance between F_α and $\mathcal{G}$ is defined as

$$(2.8) \qquad\qquad I(F_\alpha, \mathcal{G}) = I\big(F_\alpha, G_{\beta_\alpha}\big).$$

Viewed as a function of α, $I(F_\alpha, \mathcal{G})$ describes how well the various distributions of $\mathcal{F}$ are approximated by $\mathcal{G}$. Notice that $G_{\beta_\alpha} = F_\alpha$ whenever $F_\alpha \in \mathcal{G}$, in which case $I(F_\alpha, \mathcal{G}) = 0$ and β_α is called the *true value* of β relative to F_α.

Pseudo-true values play a predominant role in the theory of pseudo-ML estimation. Under additional regularity assumptions, it can be shown (Huber [1967], White [1982]) that the pseudo-ML estimator of β, based on independent observations on Y generated from F_α, converges F_α-almost surely to the pseudo-true value β_α and is asymptotically normally distributed around the pseudo-true value. These results make it clear that the pseudo-true value takes over the role of the true parameter value in the theory of ML estimation of correctly specified models.

3. Extension to conditional distributions

In order to deal with conditional distributions, the framework of Section 2 may be extended by letting the distributions of $\mathcal{F}$ and $\mathcal{G}$ depend on the value x taken by a random vector X. This random vector is defined on a probability space (C, σ_C, P), where C is the Euclidean space R^l and σ_C is the completion of the Borel σ-field on C with respect to P, the distribution of X. The distributions of $\mathcal{F}$ and $\mathcal{G}$ may now be regarded as conditional distributions, given X.

The approximation of the distributions of $\mathcal{F}$ by those of $\mathcal{G}$ may be studied under either of two viewpoints. Roughly speaking, the first approach takes into account the random nature of X, while the second approach considers the realizations of X as fixed.

Considering X as being random, it is natural to associate with each distribution $F_\alpha \in \mathcal{F}$ the distribution $G_\beta \in \mathcal{G}$ which is on average closest to F_α, where the average is taken over X. Adopting the KLIC as a criterion function then calls for minimizing $I_P(F_\alpha, G_\beta) = E_P I(F_\alpha, G_\beta)$ with respect to β. The measures ν and P, and the families of distributions $\mathcal{F}$ and $\mathcal{G}$ are now assumed to satisfy the following regularity conditions. The notation π_α refers to the product probability measure on the product σ-field in $A \times C$ generated by $\sigma_A \times \sigma_C$ which induces F_α and P. In other words, π_α is the joint distribution of Y and X, where Y is a random vector whose conditional distribution, given X, is F_α.

Assumption A1'. (a) For every $\alpha \in \Omega_\mathcal{F}$ and for P-almost all x, F_α has a conditional density f_α relative to ν, given x. (b) For P-almost all x, f_α is continuous in α.

Assumption A2'. (a) For every $\beta \in \Omega_\mathcal{G}$ and for P-almost all x, G_β has a conditional density g_β relative to ν, given x. (b) For P-almost all x, g_β is continuous in β. (c) For every $\beta \in \Omega_\mathcal{G}$ and for P-almost all x, ν is absolutely continuous with respect to G_β.

Assumption A3'. For every $\alpha \in \Omega_\mathcal{F}$, $E_P E_\alpha \log f(Y|X; \alpha)$ exists and the function $|\log g(y|x; \beta)|$ is dominated by a π_α-integrable function independent of β.

Assumption A4'. For every $\alpha \in \Omega_\mathcal{F}$, $I_P(F_\alpha, G_\beta)$ has a unique minimum with respect to $\beta \in \Omega_\mathcal{G}$.

Under Assumptions A1'–A4', it is easy to verify that the mapping

$$(3.1) \qquad \tau_{\alpha\beta} : \Omega_{\mathcal{F}} \to \Omega_{\mathcal{G}} : \alpha \mapsto \beta_\alpha,$$

where β_α is the unique solution of

$$(3.2) \qquad \max_{\beta \in \Omega_{\mathcal{G}}} E_P E_\alpha \log g(Y|X; \beta),$$

associates with each value $\alpha \in \Omega_{\mathcal{F}}$ a value $\beta_\alpha \in \Omega_{\mathcal{G}}$ such that the average distance between F_α and G_{β_α} is minimal according to the KLIC. We call β_α the *pseudo-true value* of β relative to F_α and P. Denoting the induced mapping of $\mathcal{F}$ into $\mathcal{G}$ by τ_{FG}, we call $\tau_{FG}(F_\alpha) = G_{\beta_\alpha}$ the *pseudo-true distribution* of $\mathcal{G}$ relative to F_α and P. Finally, the distance between F_α and $\mathcal{G}$ is defined as

$$(3.3) \qquad I_P(F_\alpha, \mathcal{G}) = I_P(F_\alpha, G_{\beta_\alpha}).$$

Note that β_α not only depends on F_α, but also on P. Hence, the determination of β_α requires knowledge of P unless $F_\alpha \in \mathcal{G}$, in which case $G_{\beta_\alpha} = F_\alpha$. In practice, when dealing with conditional models, P stands for the marginal distribution of the conditioning variables X, and is generally unknown. Hence, we need a notion of pseudo-true value which can be computed without knowledge of P. The following approach is then useful.

Given a vector $(x_1, \ldots, x_T)$ of independent realizations of X, one may alternatively consider the approximation of the distributions of $\mathcal{F}$ by those of $\mathcal{G}$ conditionally on these realizations. It is natural then to associate with each distribution $F_\alpha \in \mathcal{F}$ the distribution $G_\beta \in \mathcal{G}$ which is on average closest to F_α, where the average is now taken over the realizations of X. Hence, the problem (3.2) is to be replaced by

$$(3.4) \qquad \max_{\beta \in \Omega_{\mathcal{G}}} \frac{1}{T} \sum_{t=1}^{T} E_\alpha \log g(Y|x_t; \beta).$$

The solution of this problem, denoted β_α^*, minimizes the average KLIC with respect to $(x_1, \ldots, x_T)$ for a given $F_\alpha \in \mathcal{F}$. We call β_α^* the *conditional pseudo-true value* of β relative to F_α, given $(x_1, \ldots, x_T)$, or, following Gouriéroux–Monfort–Trognon [1983], the *finite sample pseudo-true value* of β relative to F_α. The conditional distance between F_α and $\mathcal{G}$, given $(x_1, \ldots, x_T)$, is defined as

$$(3.5) \qquad \bar{I}(F_\alpha, \mathcal{G}) = \frac{1}{T} \sum_{t=1}^{T} E_\alpha \log \frac{f(Y|x_t; \alpha)}{g(Y|x_t; \beta_\alpha^*)}.$$

Notice that Assumptions A1'–A4' are not sufficient either to ensure that (3.4) has a solution, or that it is unique. However, if β_α is interior to $\Omega_\mathcal{G}$, the following argument shows that, with P-probability that goes to one as $T \to \infty$, the solution exists and is unique. Given that $x_1, \ldots, x_T$ are generated independently by P, the Kolmogorov strong law of large numbers implies that

$$(3.6) \qquad \frac{1}{T} \sum_{t=1}^{T} E_\alpha \log g(Y|x_t; \beta) \xrightarrow[P]{a.s.} E_P E_\alpha \log g(Y|X; \beta),$$

where $\xrightarrow[P]{a.s.}$ denotes P-almost sure convergence as $T \to \infty$. Then, upon applying a limit result concerning the optimization of convergent random functions (White [1981], Theorem 2.1), it follows that the solution of (3.4) converges P-almost surely to the solution of (3.2). This establishes the asymptotic existence and uniqueness of β_α^* as well as the asymptotic equivalence of β_α and β_α^*.

4. Pseudo-true values of location and scale parameters

Many commonly used parametric families of distributions have a location and/or a scale parameter. In such cases, the solution of (2.6) exhibits some invariances, which may be exploited to reduce the calculations needed to determine the pseudo-true value. We confine ourselves to unconditional distributions, although similar results hold in the more general context of conditional distributions—see Broze–Gouriéroux [1993], who deduce consistency results from such invariances. To fix the ideas, consider the measure space $(R, \mathcal{B}, \lambda)$, where $\mathcal{B}$ is the Borel σ-field on R and λ denotes the Lebesgue measure, and assume that $\mathcal{F}$ and $\mathcal{G}$ are families of continuous univariate distributions, parametrized such that

$$(4.1) \qquad \alpha = (\alpha_1, \alpha_2) \in R \times R_{++},$$

$$(4.2) \qquad \beta = (\beta_1, \beta_2) \in R \times R_{++},$$

and with density functions f_α and g_β satisfying

$$(4.3) \qquad f(y; \alpha) = \frac{1}{\alpha_2}\, r\!\left(\frac{y - \alpha_1}{\alpha_2}\right),$$

$$(4.4) \qquad g(y; \beta) = \frac{1}{\beta_2}\, s\!\left(\frac{y - \beta_1}{\beta_2}\right),$$

for some functions r and s. Then, α_1 and β_1 are location parameters, and α_2 and β_2 are scale parameters. The following lemma characterizes the functional dependence of β_α on α.

Lemma 4.1. *Let Assumptions A1–A5 hold, and let $\mathcal{F}$ and $\mathcal{G}$ be families of continuous univariate distributions with parametrizations satisfying (4.1)– (4.2) and density functions satisfying (4.3)–(4.4). Then, for some constants c_0, c_1 and c_2 not depending on α,*

$$(4.5) \qquad \beta_\alpha = (\alpha_1 + c_1\alpha_2, c_2\alpha_2)$$

and

$$(4.6) \qquad I(F_\alpha, \mathcal{G}) = c_0.$$

Furthermore, $c_1 = 0$ when the density functions $f(y; \alpha)$ and $g(y; \beta)$ are symmetric around $y = \alpha_1$ and $y = \beta_1$, respectively.

Proof. From (4.3)–(4.4) and definition (2.1), it follows that

$$(4.7) \qquad I(F_\alpha, G_\beta) = \int_{S_\alpha} \log\left[\frac{\beta_2\, r\!\left(\frac{y - \alpha_1}{\alpha_2}\right)}{\alpha_2\, s\!\left(\frac{y - \beta_1}{\beta_2}\right)}\right] \frac{1}{\alpha_2}\, r\!\left(\frac{y - \alpha_1}{\alpha_2}\right) dy,$$

where S_α is the support of f_α, i.e., S_α is the closure of $\{y \mid f_\alpha(y) > 0\}$. Upon substituting z for $(y - \alpha_1)/\alpha_2$ under the integral, (4.7) can be rewritten as

$$(4.8) \qquad I(F_\alpha, G_\beta) = \int_{S_r} \log\left[\frac{\gamma_2\, r(z)}{s\!\left(\frac{z - \gamma_1}{\gamma_2}\right)}\right] r(z)dz,$$

where S_r is the closure of $\{y \mid r(y) > 0\}$ and $\gamma = (\gamma_1, \gamma_2)$ is defined as

$$(4.9) \qquad \gamma = \left(\frac{\beta_1 - \alpha_1}{\alpha_2}, \frac{\beta_2}{\alpha_2}\right).$$

15

Notice that, for every $\alpha \in \Omega_{\mathcal{F}}$, the transformation from β to γ is monotonically increasing in each component. Therefore, minimizing $I(F_\alpha, G_\beta)$ with respect to β is equivalent to minimizing $I(F_\alpha, G_\beta)$ with respect to γ. By assumption, $I(F_\alpha, G_\beta)$ attains a unique minimum, say at $\gamma = (c_1, c_2)$. From (4.8) we see that c_1 and c_2 do not depend on α. Transforming this solution back to β by means of (4.9) results in (4.5). Furthermore, $I(F_\alpha, \mathcal{G})$ is equal to the right hand side (RHS) of (4.8) evaluated at $\gamma = (c_1, c_2)$, wherefrom (4.6) follows.

If additional symmetry assumptions are made with respect to $f(y; \alpha)$ and $g(y; \beta)$, the functions r and s are even, i.e., $r(-y) = r(y)$ and $s(-y) = s(y)$. Then, denoting the RHS of (4.8) by $L(\gamma_1, \gamma_2)$, observe that $L(\gamma_1, \gamma_2) = L(-\gamma_1, \gamma_2)$, upon substituting $-z$ for z under the integral. Thus, if L has a minimum at (c_1, c_2), it also has a minimum at $(-c_1, c_2)$. Since the minimum is unique by assumption, it follows that $c_1 = 0$. □

Under the conditions of the above lemma, it follows that if (c_1, c_2) is the pseudo-true value of β relative to $F = F_\alpha|_{\alpha=(0,1)}$, then $(\alpha_1 + c_1\alpha_2, c_2\alpha_2)$ is the pseudo-true value of β relative to F_α. It is important to notice that this device reduces a problem with a functional solution β_α to a problem with a numerical solution (c_1, c_2). If no closed form solution for β_α exists, it is always possible to determine (c_1, c_2) numerically up to the required accuracy and then to use (4.5). If moreover the density functions are symmetric around their location parameters, then $c_1 = 0$ and one only has to find c_2 as the unique solution of

$$(4.10) \qquad \min_{\gamma_2 \in R_{++}} \left[-H(F) + \log \gamma_2 - \int_{S_r} \log\left[s(z/\gamma_2)\right] r(z) dz \right]$$

—see (4.8). From the first-order condition corresponding to this minimum, it follows that c_2 has to be solved from

$$(4.11) \qquad c_2 + E_F \left[\frac{s'(Y/c_2)}{s(Y/c_2)} Y \right] = 0,$$

where Y is a random variable whose distribution is F, and s' is the first-order derivative of s. Finally, a general formula for $I(F_\alpha, \mathcal{G})$ follows from (4.8) as

$$(4.12) \qquad I(F_\alpha, \mathcal{G}) = -H(F) + \log c_2 - E_F \log s\left(\frac{Y - c_1}{c_2}\right).$$

16

Lemma 4.1 can be specialized and extended in many directions. We will not give an all-embracing formulation of it because its statement would be too involved and the necessary notation would obscure the essentials. Rather, we will discuss in a fairly loose way how modifications of the conditions of the lemma affect the results. Proofs may be found along the same lines as above.

Upon introducing matrix notation, the lemma generalizes immediately to the case where $\mathcal{F}$ and $\mathcal{G}$ are families of multivariate distributions, with location vectors α_1 and β_1 and scaling matrices α_2 and β_2. In (4.5), c_1 is then a (row) vector of constants, and c_2 a square matrix of constants. Also, (4.6) continues to hold.

Besides having a location and a scale parameter, the distributions of $\mathcal{F}$ and $\mathcal{G}$ may also depend on a third parameter (vector), say α_3 and β_3. Then, $\beta_\alpha = (\alpha_1 + c_1\alpha_2, c_2\alpha_2, c_3)$ obtains as the extension of (4.5), where c_1, c_2 and c_3 are now functions of α_3. Also, c_0 becomes a function of α_3.

It is also of interest to consider the cases where either one or both of $\mathcal{F}$ and $\mathcal{G}$ have only a location or a scale parameter. The absence of a location (resp. scale) parameter may always be conceived as fixing the location (resp. scale) parameter at zero (resp. one), perhaps at the cost of a reparametrization. If $\mathcal{G}$ has only a location parameter, then $\beta_\alpha = \alpha_1 + c_1(\alpha_2)\alpha_2$ and $I(F_\alpha, \mathcal{G}) = c_0(\alpha_2)$, where c_0 and c_1 are functions of α_2 only. If $\mathcal{G}$ has only a scale parameter, then $\beta_\alpha = c_2(\alpha_1/\alpha_2)\alpha_2$ and $I(F_\alpha, \mathcal{G}) = c_0(\alpha_1/\alpha_2)$, where c_0 and c_2 are functions of α_1/α_2 only. The results for the cases where $\mathcal{F}$ has only a location or a scale parameter are obtained immediately, by letting $\alpha_1 = 0$ or $\alpha_2 = 1$.

It may also be observed that $c_1 = 0$ under more general conditions than are given by the lemma, provided that the density functions are symmetric around their location parameter (or around zero in the absence thereof).

Finally, by choosing an appropriate measure ν instead of the Lebesgue measure λ, the result of the lemma carries over to families of distributions of any type, as long as they satisfy (4.1)–(4.4).

5. Examples

In this section we derive pairwise pseudo-true values for some classes of commonly used parametric families of distributions. The first example deals with some families of symmetric distributions on the line, namely the families of rectangular, normal, Laplace and logistic distributions. It also provides an application of Lemma 4.1. In Example 5.2 we consider two families of distributions on the positive half-line, namely the gamma and the lognormal families, and we illustrate an extension to Lemma 4.1. The last two examples are devoted to conditional distributions. Example 5.3 studies univariate normal linear regression models which assume the same dependent variable. This case, sometimes referred to as the choice of regressors problem, has received considerable attention in the literature—see Pesaran [1974], Amemiya [1980], Gouriéroux–Monfort–Trognon [1983], Mizon [1984], Mizon–Richard [1986], Hendry–Richard [1990] and Gouriéroux–Monfort [1995], among others. In Example 5.4, we consider a univariate normal regression model and a univariate lognormal regression model. The choice between normal and lognormal models has also received much attention in the literature—see, e.g., Cox [1962], Sargan [1964], Aneuryn-Evans–Deaton [1980], Gouriéroux–Monfort–Trognon [1983] and Hendry–Richard [1990]. Note, however, that we are only interested in the pseudo-true values in this chapter. Considerations dealing with model specification, hypothesis testing and model choice will be given in later chapters.

Example 5.1. Consider the measure space $(R, \mathcal{B}, \lambda)$, where λ denotes the Lebesgue measure. Denote the families of (non-degenerate) rectangular, normal, Laplace and logistic distributions on $(R, \mathcal{B})$ by $\mathcal{F} = \{F_\alpha | \alpha \in \Omega_\mathcal{F}\}$, $\mathcal{G} = \{G_\beta | \beta \in \Omega_\mathcal{G}\}$, $\mathcal{P} = \{P_\gamma | \gamma \in \Omega_\mathcal{P}\}$ and $\mathcal{Q} = \{Q_\delta | \delta \in \Omega_\mathcal{Q}\}$, respectively. See, for example, Johnson–Kotz [1970a, 1970b] for a discussion of these distributions. The families can be parametrized in such a way that the first component of the parameter vector is a location parameter and the second component a scale parameter whose square is equal to the variance of the distribution. With these parametrizations, all parameter spaces are equal to $R \times R_{++}$, and the respective density functions are given by

$$(5.1) \qquad f(y; \alpha) = \frac{1}{2\sqrt{3}\alpha_2} 1_{[-1,1]} \left(\frac{y - \alpha_1}{\sqrt{3}\alpha_2} \right),$$

$$(5.2) \qquad g(y;\beta) = \frac{1}{\sqrt{2\pi\beta_2}} \exp\left[-\frac{(y-\beta_1)^2}{2\beta_2^2}\right],$$

$$(5.3) \qquad p(y;\gamma) = \frac{1}{\sqrt{2\gamma_2}} \exp\left[-\frac{\sqrt{2}\,|y-\gamma_1|}{\gamma_2}\right],$$

$$(5.4) \qquad q(y;\delta) = \frac{\pi \exp\left(-\frac{\pi(y-\delta_1)}{\sqrt{3}\delta_2}\right)}{\sqrt{3}\delta_2\left[1+\exp\left(-\frac{\pi(y-\delta_1)}{\sqrt{3}\delta_2}\right)\right]^2},$$

where $1_S(\cdot)$ is the indicator function of the set S. One could now, for each pair of $\mathcal{F}, \mathcal{G}, \mathcal{P}$ and $\mathcal{Q}$, calculate the pseudo-true values by solving a problem like (2.6). Furthermore, the distance between any given distribution and any given family can be calculated by the use of (2.8). However, given the parametrizations and the symmetry of the distributions involved, we may straightforwardly apply Lemma 4.1. Thus, we only need to find c_2 as the solution of (4.11) and subsequently calculate $I(\cdot,\cdot)$ from (4.12) with $c_1 = 0$.

Tables 1 and 2 give the results of these calculations, which are carried out in more detail in Appendix A. The appendix also gives the exact numbers whenever we could find them in terms of standard mathematical functions.

The diagonals of the tables are self-evident. Note that the pseudo-true values of α relative to G_β, P_γ and Q_δ do not exist. This is a consequence of the fact that the Lebesgue measure is not absolutely continuous with respect to F_α so that F_α violates Assumption A2-(c) (where G_β is taken to be F_α). The problem may be understood more directly by noticing that the support of f_α, which is the interval $[\alpha_1 - \sqrt{3}\alpha_2, \alpha_1 + \sqrt{3}\alpha_2]$, never contains the support of any of g_β, p_γ and q_δ, which is the real line. Since, however, all distributions have a density relative to a dominating measure (i.e., λ), it makes sense to allow for an infinite value of the KLIC—see the first column of Table 2. Note furthermore that the families of normal, Laplace and logistic distributions do not provide close approximations to the rectangular distribution, as indicated by Table 2. Of course, this is hardly surprising given the shape of the distributions involved. The figures in the table also reflect the well-known feature that the normal and the logistic distributions are close approximations to each other.

19

Table 1. *Pairwise pseudo-true scale parameters relative to rectangular ($F \in \mathcal{F}$), normal ($G \in \mathcal{G}$), Laplace ($P \in \mathcal{P}$) and logistic ($Q \in \mathcal{Q}$) distributions.**

	$\mathcal{F}$	$\mathcal{G}$	$\mathcal{P}$	$\mathcal{Q}$
F	1	1	1.2247	1.1024
G	–	1	1.1284	1.0371
P	–	1	1	0.9446
Q	–	1	1.0809	1

* The entry (i,j) is the pseudo-true value of the scale parameter of the family j relative to the distribution i.

Table 2. *Pairwise distances between rectangular ($\mathcal{F}$), normal ($\mathcal{G}$), Laplace ($\mathcal{P}$) and logistic ($\mathcal{Q}$) families.**

	$\mathcal{F}$	$\mathcal{G}$	$\mathcal{P}$	$\mathcal{Q}$
F	0	0.1765	0.3069	0.2217
G	$+\infty$	0	0.0484	0.0095
P	$+\infty$	0.0724	0	0.0219
Q	$+\infty$	0.0144	0.0198	0

* The entry (i,j) is the distance between distribution i and family j.

The second column of Table 1 could have been anticipated from the fact that the pseudo-ML estimator of the scale of the normal distribution equals the square root of the empirical variance, which, given the parametrizations, converges almost surely to one. We see that the other pseudo-true scales may differ substantially from one. However, one should not be misled and draw the conclusion that the normal family is in some sense superior to the other ones. In fact, reparametrizing the families such that the scale is equal to the mean absolute deviation of the distribution would make the third column of Table 1 identically equal to one, since the pseudo-ML estimator of the scale of the Laplace distribution equals the empirical mean absolute deviation. On the other hand, the pseudo-ML estimator of the scale of the logistic distribution does not possess a closed form. As a consequence, the reparametrizations of the families that would make the fourth column of Table 1 identically equal to one would involve integrals that do not have solutions in terms of standard mathematical functions. For the same reason, the fourth column of Tables 1 and 2 had to be determined numerically instead of analytically—see Appendix A.

Figure 1 depicts the density functions associated with the pseudo-true distributions whenever these exist. UL, UR, etc. refer to the upper-left, upper-right, etc. parts of the figure. The figure is an illustration of the conclusions that can be drawn from Table 2. It also provides a quantitative insight into the absolute order of magnitude of the distances as measured by the KLIC. For example, observe from the UR part of the figure that the difference between the normal distribution and its logistic approximation is non-negligible, despite the rather small figure in Table 2 corresponding to this approximation.

Figure 1. *Pseudo-true densities relative to rectangular (UL), normal (UR), Laplace (LL) and logistic (LR) distributions.**

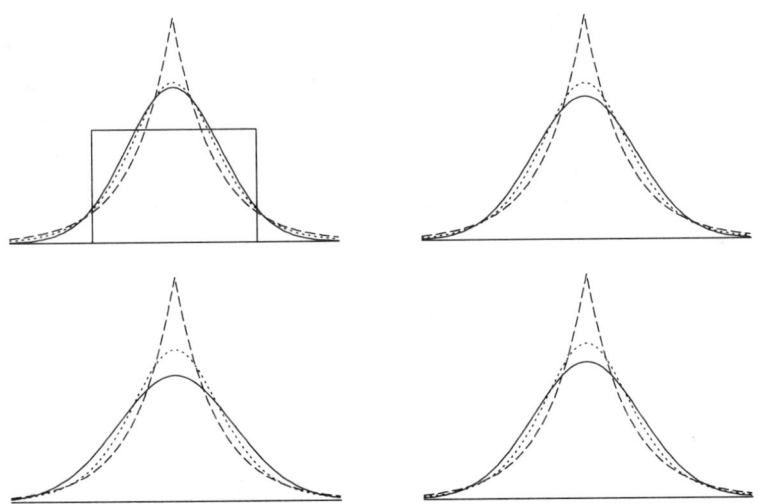

* Solid lines: rectangular and normal densities; dashed line: Laplace density; dotted line: logistic density.   □

Example 5.2. Consider the measure space $(R, \mathcal{B}, \nu)$, where $\nu(B) = \lambda(B \cap R_+)$ for every $B \in \mathcal{B}$ and λ denotes the Lebesgue measure. Denote the families of gamma and lognormal distributions on $(R, \mathcal{B})$ by $\mathcal{F} = \{F_\alpha | \alpha \in \Omega_{\mathcal{F}}\}$ and $\mathcal{G} = \{G_\beta | \beta \in \Omega_{\mathcal{G}}\}$, respectively. A discussion of these distributions may be found in Johnson–Kotz [1970a]. We choose the parametrizations of

$\mathcal{F}$ and $\mathcal{G}$ leading to the following density functions ($y > 0$):

$$(5.5) \qquad f(y; \alpha) = \frac{1}{\Gamma(\alpha_1)\alpha_2^{\alpha_1}} y^{\alpha_1 - 1} \exp\left(-\frac{y}{\alpha_2}\right),$$

$$(5.6) \qquad g(y; \beta) = \frac{1}{\sqrt{2\pi\beta_1}\, y} \exp\left[-\frac{(\log y - \log \beta_2)^2}{2\beta_1}\right],$$

where $\Gamma(\cdot)$ is the gamma function—see Appendix B. The parameter spaces are $\Omega_{\mathcal{F}} = \Omega_{\mathcal{G}} = R_{++}^2$. Although obviously arbitrary, the parametrization of $\mathcal{G}$ is somewhat unusual. G_β is the distribution of a random variable whose logarithm is normally distributed with mean $\log \beta_2$ and variance β_1. This parameter choice is motivated by the form of the results which follow. The invariance property of the pseudo-true value under reparametrizations allows the results to be transformed to other parametrizations. Upon inspecting the density functions, we see that α_2 and β_2 are scale parameters. The pseudo-true values relative to F_α or G_β may therefore by found by first calculating the pseudo-true values relative to $F = F_\alpha|_{\alpha_2=1}$ or $G = G_\beta|_{\beta_2=1}$ and then applying an appropriate extension of Lemma 4.1. Note the dependence of F on α_1 and of G on β_1. A similar short-cut applies to the calculation of the distances. Appendix B gives the details of the calculations. The resulting pseudo-true values are

$$(5.7) \qquad \beta_\alpha = \left(\psi'(\alpha_1), \exp(\psi(\alpha_1))\alpha_2\right),$$

$$(5.8) \qquad \alpha_\beta = \left(\kappa(\beta_1), \exp\left(-\psi(\kappa(\beta_1))\right)\beta_2\right),$$

where $\psi(\cdot)$ and $\psi'(\cdot)$ are the digamma and trigamma functions, and $\kappa(\beta_1)$ is the functional solution of

$$(5.9) \qquad \log(\kappa) - \psi(\kappa) - \frac{1}{2}\beta_1 = 0.$$

The distances are given by

$$(5.10) \quad I(F_\alpha, \mathcal{G}) = \frac{1}{2}\log\psi'(\alpha_1) - \log\Gamma(\alpha_1) - \alpha_1[1 - \psi(\alpha_1)] + \frac{1}{2}\log(2\pi) + \frac{1}{2},$$

$$(5.11) \quad I(G_\beta, \mathcal{F}) = \log\Gamma(\kappa(\beta_1)) + \kappa(\beta_1)[1 - \psi(\kappa(\beta_1))] - \frac{1}{2}\log(2\pi\beta_1) - \frac{1}{2}.$$

As a result of the invariances, these distances depend only on the non-scale parameters α_1 and β_1. The dependencies of $I(F_\alpha, \mathcal{G})$ on α_1 and of $I(G_\beta, \mathcal{F})$ on β_1 are shown in Figure 2. Motivated by the particular functional form and the analytical results which follow, we set out $I(G_\beta, \mathcal{F})$ as a function of β_1^{-1}.

Figure 2. *Distances between the gamma distribution and the lognormal family and between the lognormal distribution and the gamma family.** *

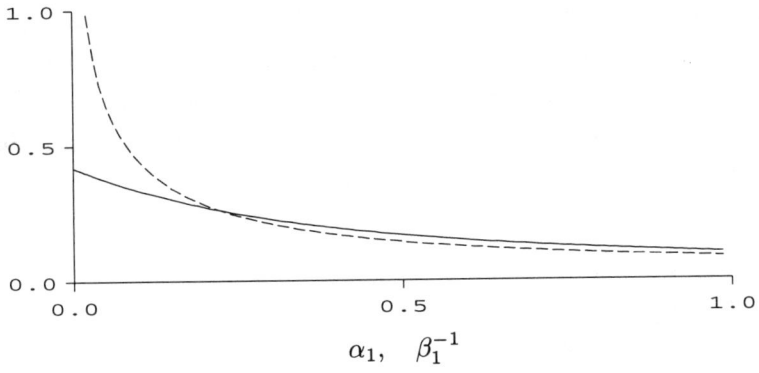

* Solid line: $I(F_\alpha, \mathcal{G})$, distance between gamma distribution and lognormal family; dashed line: $I(G_\beta, \mathcal{F})$, distance between lognormal distribution and gamma family.

In Appendix B, we show that

(5.12) $$\lim_{\alpha_1 \to 0} I(F_\alpha, \mathcal{G}) = \frac{1}{2} \log(2\pi) - \frac{1}{2} = 0.4189,$$

(5.13) $$\lim_{\alpha_1 \to +\infty} I(F_\alpha, \mathcal{G}) = 0,$$

and that

(5.14) $$\lim_{\beta_1 \to 0} I(G_\beta, \mathcal{F}) = 0,$$

(5.15) $$\lim_{\beta_1 \to +\infty} I(G_\beta, \mathcal{F}) = +\infty.$$

The distance between the gamma distribution and the lognormal family is bounded above by 0.4189. By contrast, the distance between the lognormal distribution and the gamma family is unbounded as $\beta_1 \to +\infty$. The intuition behind (5.13) is that, as $\alpha_1 \to +\infty$, the gamma distribution and the

23

associated pseudo-true lognormal distribution converge to each other, as we will prove. The gamma distribution, having mean $\alpha_1\alpha_2$ and variance $\alpha_1\alpha_2^2$, shifts to the right and becomes flatter as $\alpha_1 \to \infty$. Therefore, the convergence condition is formulated in terms of a standardized random variable, viz.

$$(5.16) \qquad \lim_{\alpha_1 \to +\infty} \mathrm{Pr}_\alpha \left[\frac{Y - \alpha_1\alpha_2}{\sqrt{\alpha_1\alpha_2}} < a \right] = \lim_{\alpha_1 \to +\infty} \mathrm{Pr}_{\beta_\alpha} \left[\frac{Y - \alpha_1\alpha_2}{\sqrt{\alpha_1\alpha_2}} < a \right]$$

for all $a \in R$. Pr_α and $\mathrm{Pr}_{\beta_\alpha}$ denote probabilities under F_α and G_{β_α}, respectively. We prove in Appendix B that both sides of (5.16) are equal to $\Phi(a)$, where $\Phi(\cdot)$ is the standard normal distribution function. It follows that, as $\alpha_1 \to +\infty$, the standardized gamma distribution and the standardized pseudo-true lognormal distribution both converge to the standard normal distribution. A similar argument goes with (5.14). □

Example 5.3. Consider the measure space $(R, \mathcal{B}, \lambda)$, where λ denotes the Lebesgue measure. Let X be a random vector defined on the probability space (C, σ_C, P), where C is the Euclidean space R^l and P is the distribution of X. Let X_F and X_G be σ_C-measurable functions of X, with values in R^m and R^n, respectively, and denote the values of these functions for a realization x of X by x_F and x_G, respectively. X_F and X_G may be subvectors of X, but more complicated functions of X are equally well allowed. In particular, l imposes no constraints on m and n. Consider the family $\mathcal{F} = \{F_\alpha | \alpha \in \Omega_{\mathcal{F}}\}$, where F_α is the conditional normal distribution on $(R, \mathcal{B})$, given X, with conditional mean $X_F'\alpha_1$ and conditional variance α_2, and where $\Omega_{\mathcal{F}} = R^m \times R_{++}$. Consider also $\mathcal{G} = \{G_\beta | \beta \in \Omega_{\mathcal{G}}\}$, where G_β is the conditional normal distribution on $(R, \mathcal{B})$, given X, with conditional mean $X_G'\beta_1$ and conditional variance β_2, and where $\Omega_{\mathcal{G}} = R^n \times R_{++}$. The conditional density functions of F_α and G_β are given by

$$(5.17) \qquad f(y|x; \alpha) = \frac{1}{\sqrt{2\pi\alpha_2}} \exp\left[-\frac{(y - x_F'\alpha_1)^2}{2\alpha_2} \right],$$

and

$$(5.18) \qquad g(y|x; \beta) = \frac{1}{\sqrt{2\pi\beta_2}} \exp\left[-\frac{(y - x_G'\beta_1)^2}{2\beta_2} \right],$$

respectively.

We shall now calculate the pseudo-true value of β relative to F_α and P. Taking expectations of the log-densities relative to F_α and P, we have

$$(5.19) \qquad E_P E_\alpha \log f(Y|X;\alpha) = -\frac{1}{2}\log(2\pi) - \frac{1}{2}\log\alpha_2 - \frac{1}{2}$$

and

$$
\begin{aligned}
E_P E_\alpha \log g(Y|X;\beta) &= -\frac{1}{2}\log(2\pi) - \frac{1}{2}\log\beta_2 - E_P E_\alpha \frac{(Y - X'_G\beta_1)^2}{2\beta_2}, \\
(5.20) \qquad &= -\frac{1}{2}\log(2\pi) - \frac{1}{2}\log\beta_2 - \frac{\alpha_2 + E_P(m_1 - X'_G\beta_1)^2}{2\beta_2},
\end{aligned}
$$

where $m_1 = X'_F\alpha_1$, the conditional mean of F_α. Recall from (3.2) that the pseudo-true value of β relative to F_α and P maximizes $E_P E_\alpha \log g(Y|X;\beta)$ with respect to β. The solution of this problem is given by

$$(5.21) \quad \beta_\alpha = \begin{pmatrix} [E_P(X_G X'_G)]^{-1} E_P(X_G m_1) \\ \alpha_2 + E_P(m_1^2) - E_P(m_1 X'_G) [E_P(X_G X'_G)]^{-1} E_P(X_G m_1) \end{pmatrix}.$$

The existence of a maximum requires the existence of the expectations of m_1^2, $m_1 X'_G$ and $X_G X'_G$ relative to P. Thus m_1 and the components of X_G need to be in the linear space $L_2(R^l, P)$ of real-valued functions on R^l which are square integrable with respect to P. The maximum is unique if, moreover, $E_P(X_G X'_G)$ is non-singular. (If $E_P(X_G X'_G)$ is singular, the solutions of (3.2) are still given by (5.21) if we replace the inverse by any generalized inverse.) Notice that the conditional variance of G_{β_α} is larger than or equal to the conditional variance of F_α. The average distance between F_α and $\mathcal{G}$ follows upon subtracting (5.20), with $\beta = \beta_\alpha$, from (5.19):

$$(5.22) \qquad I_P(F_\alpha, \mathcal{G}) = \frac{1}{2}\log\left(\frac{\beta_{2\alpha}}{\alpha_2}\right),$$

where $\beta_{2\alpha}$ is the conditional variance of G_{β_α}. Note that α_β and $I_P(G_\beta, \mathcal{F})$ follow from (5.21) and (5.22) by reversing the roles of $\mathcal{F}$ and $\mathcal{G}$.

Since $L_2(R^l, P)$, equipped with the norm

$$(5.23) \qquad \|h\| = (E_P h^2)^{1/2}, \qquad h \in L_2(R^l, P),$$

and the inner product

$$(5.24) \qquad (h_1, h_2) = E_P(h_1 h_2), \qquad h_1, h_2 \in L_2(R^l, P),$$

is complete and is therefore a Hilbert space, we can write $L_2(R^l, P)$ as the direct sum of any given linear subspace of $L_2(R^l, P)$ and its orthogonal complement. In turn, this allows any given point of $L_2(R^l, P)$ to be written as the sum of its orthogonal projections onto these orthogonal subspaces. Now in view of (5.21), the conditional mean and variance of G_{β_α} have interpretations in terms of orthogonal projections in $L_2(R^l, P)$. Let $p_1 = X_G'[E_P(X_G X_G')]^{-1} E_P(X_G m_1)$ be the orthogonal projection of m_1 onto the linear subspace of $L_2(R^l, P)$ spanned by the components of X_G and $p_2 = m_1 - p_1$ the orthogonal projection of m_1 onto the orthogonal complement of X_G. Then we see from (5.21) that the conditional mean of G_{β_α} equals p_1 and that the conditional variance of G_{β_α} exceeds the conditional variance of F_α by the square norm of p_2.

Given a vector $(x_1, \ldots, x_T)$ of realizations of X, we now turn to the conditional pseudo-true value of β relative to F_α, given $(x_1, \ldots, x_T)$. Denote the $T \times m$ and $T \times n$ matrices $\big(x_F(x_1), \ldots, x_F(x_T)\big)'$ and $\big(x_G(x_1), \ldots, x_G(x_T)\big)'$ by Ξ_F and Ξ_G, respectively. Taking averages over the realizations $x_1, \ldots, x_T$ of the expectations of the log-densities relative to F_α, we have

$$(5.25) \qquad \frac{1}{T} \sum_{t=1}^{T} E_\alpha \log f(Y|x_t; \alpha) = -\frac{1}{2}\log(2\pi) - \frac{1}{2}\log\alpha_2 - \frac{1}{2}$$

and

$$\frac{1}{T} \sum_{t=1}^{T} E_\alpha \log g(Y|x_t; \beta) = -\frac{1}{2}\log(2\pi) - \frac{1}{2}\log\beta_2$$

$$(5.26) \qquad\qquad -\frac{\alpha_2 + \frac{1}{T}(\mu_1 - \Xi_G\beta_1)'(\mu_1 - \Xi_G\beta_1)}{2\beta_2},$$

where $\mu_1 = \Xi_F\alpha_1$. According to (3.4), the conditional pseudo-true value β_α^* maximizes the RHS of (5.26) with respect to β, and is thus given by

$$(5.27) \qquad \beta_\alpha^* = \begin{pmatrix} (\Xi_G'\Xi_G)^{-1}\Xi_G'\mu_1 \\ \alpha_2 + \frac{1}{T}\mu_1'\left[I - \Xi_G(\Xi_G'\Xi_G)^{-1}\Xi_G'\right]\mu_1 \end{pmatrix},$$

provided that Ξ_G has full column rank, which requires $T \geq n$. (If the rank of Ξ_G is less than n, then (3.4) has multiple solutions, which are given by (5.27) with the inverse replaced by any generalized inverse.) Subtracting (5.26), with $\beta = \beta_\alpha^*$, from (5.25) gives the average distance between F_α and $\mathcal{G}$ as

$$(5.28) \qquad\qquad \bar{I}(F_\alpha, \mathcal{G}) = \frac{1}{2}\log\left(\frac{\beta_{2\alpha}^*}{\alpha_2}\right),$$

where $\beta_{2\alpha}^*$ is the conditional variance of $G_{\beta_\alpha^*}$. Again, α_β^* and $\bar{I}(G_\beta, \mathcal{F})$ follow from (5.27) and (5.28) by reversing the roles of $\mathcal{F}$ and $\mathcal{G}$.

The conditional mean and variance of the pseudo-true distribution now have interpretations in terms of orthogonal projections in the Euclidean space R^T, equipped with the usual norm and inner product. Let F_α^* (resp. G_β^*) be the product probability measure induced by the measures F_α (resp. G_β), given $X = x_t$, $t = 1, \ldots, T$. Then, F_α^* (resp. G_β^*) is the conditional normal distribution with mean μ_1 (resp. $\Xi_G \beta_1$) and conditional covariance matrix $\alpha_2 I$ (resp. $\beta_2 I$). Furthermore, let $p_1 = \Xi_G (\Xi_G' \Xi_G)^{-1} \Xi_G' \mu_1$ be the orthogonal projection of μ_1 onto the linear subspace spanned by the columns of Ξ_G and $p_2 = \mu_1 - p_1$ the orthogonal projection of μ_1 onto the orthogonal complement of the columns of Ξ_G. Then we see from (5.27) that the conditional mean of $G_{\beta_\alpha^*}^*$ equals p_1 and that the conditional variance of $G_{\beta_\alpha^*}^*$ exceeds the conditional variance of F_α^* by the square norm of p_2, divided by T.

Clearly, the expectations relative to P in (5.19)–(5.22) are replaced by averages over the realizations of X in (5.25)–(5.27). This is the distinguishing feature between the unconditional and the conditional pseudo-true values and distances. $\qquad\qquad\qquad\qquad\qquad\qquad\qquad\qquad\qquad\qquad\qquad$ □

Example 5.4. Consider the setting of Example 5.3, but let $\mathcal{F} = \{F_\alpha | \alpha \in \Omega_\mathcal{F}\}$ now be the family of conditional lognormal distributions on $(R, \mathcal{B})$, given X, with conditional log-mean $X_F' \alpha_1$ and conditional log-variance α_2. As before, $\Omega_\mathcal{F} = R^m \times R_{++}$. The conditional density function of F_α is given by

$$(5.29) \qquad f(y|x; \alpha) = \frac{1}{\sqrt{2\pi\alpha_2}y} \exp\left[-\frac{(\log y - x_F'\alpha_1)^2}{2\alpha_2} \right].$$

With $\mathcal{G}$ defined as in Example 5.3, it follows that, for P-almost all X, F_α is absolutely continuous with respect to λ, G_β and λ are absolutely continuous with respect to each other, and G_β is not absolutely continuous with respect to F_α. As a result, neither the pseudo-true value of α relative to G_β and P, nor the conditional pseudo-true value of α relative to G_β, given $(x_1, \ldots, x_T)$, exists. Thus, we only have to determine the unconditional and conditional pseudo-true values of β relative to F_α.

It follows from properties of the lognormal distribution that the conditional mean and variance of F_α, given X, are given by

$$(5.30) \qquad\qquad m_1 = \exp(X_F'\alpha_1 + \tfrac{1}{2}\alpha_2)$$

and

$$(5.31) \qquad\qquad m_2 = \exp(2X_F'\alpha_1 + 2\alpha_2) - m_1^2,$$

respectively. Using (5.30)–(5.31), we have

$$(5.32) \quad E_P E_\alpha \log f(Y|X;\alpha) = -\frac{1}{2}\log(2\pi) - \frac{1}{2}\log\alpha_2 - E_P(X_F'\alpha_1) - \frac{1}{2}$$

and

(5.33)

$$E_P E_\alpha \log g(Y|X;\beta) = -\frac{1}{2}\log(2\pi) - \frac{1}{2}\log\beta_2 - \frac{E_P m_2 + E_P(m_1 - X_G'\beta_1)^2}{2\beta_2}.$$

The pseudo-true value of β relative to F_α and P maximizes the RHS of (5.33) with respect to β and follows easily as

(5.34)

$$\beta_\alpha = \begin{pmatrix} [E_P(X_G X_G')]^{-1} E_P(X_G m_1) \\ E_P m_2 + E_P(m_1^2) - E_P(m_1 X_G') [E_P(X_G X_G')]^{-1} E_P(X_G m_1) \end{pmatrix}.$$

A necessary condition for the existence of a maximum is that $\exp(X_F'\alpha_1)$ and the components of X_G are functions of $L_2(R^l, P)$. Observe from (5.30) and (5.31) that, under this condition, $m_1 \in L_2(R^l, P)$ and $E_P m_2$ exists. The maximum is unique if and only if, in addition, $E_P(X_G X_G')$ is non-singular. The average distance between F_α and $\mathcal{G}$ follows as the difference between (5.32) and (5.33), with $\beta = \beta_\alpha$,

$$(5.35) \qquad\qquad I_P(F_\alpha, \mathcal{G}) = \frac{1}{2}\log\left(\frac{\beta_{2\alpha}}{\alpha_2}\right) - E_P(X_F'\alpha_1),$$

where $\beta_{2\alpha}$ is the conditional variance of $\mathcal{G}_{\beta_\alpha}$. A necessary and sufficient condition for the distance to be finite is that $X_F'\alpha_1$ be P-integrable.

The conditional mean and variance of $\mathcal{G}_{\beta_\alpha}$ have interpretations in terms of orthogonal projections in $L_2(R^l, P)$ which are very similar to those in Example 5.3. The only difference lies in the fact that the conditional variance of F_α now depends on X, so that it is the mean conditional variance $E_P(m_2)$ instead of α_2 which determines the conditional variance of G_{β_α}.

Given a vector $(x_1, \ldots, x_T)$ of realizations of X, denote the $T \times m$ and $T \times n$ matrices $(x_F(x_1), \ldots, x_F(x_T))'$ and $(x_G(x_1), \ldots, x_G(x_T))'$ by Ξ_F and Ξ_G, respectively. Furthermore, define the $T \times 1$ vectors

$$(5.36) \qquad \mu_1 = \left[\exp\left(x_F'(x_t)\alpha_1 + \tfrac{1}{2}\alpha_2\right)\right]$$

and

$$(5.37) \qquad \mu_2 = \left[\exp\left(2x_F'(x_t)\alpha_1 + 2\alpha_2\right) - \exp\left(2x_F'(x_t)\alpha_1 + \alpha_2\right)\right],$$

which are the vector analogues to (5.30) and (5.31). Taking averages over the realizations $x_1, \ldots, x_T$ of the expectations of the log-densities relative to F_α, we have

$$(5.38) \qquad \frac{1}{T}\sum_{t=1}^{T} E_\alpha \log f(Y|x_t; \alpha) = -\frac{1}{2}\log(2\pi) - \frac{1}{2}\log\alpha_2 - \frac{1}{T}\iota'\Xi_F\alpha_1 - \frac{1}{2}$$

and

$$\frac{1}{T}\sum_{t=1}^{T} E_\alpha \log g(Y|x_t; \beta) = -\frac{1}{2}\log(2\pi) - \frac{1}{2}\log\beta_2$$

$$(5.39) \qquad \qquad -\frac{1}{T}\frac{\iota'\mu_2 + (\mu_1 - \Xi_G\beta_1)'(\mu_1 - \Xi_G\beta_1)}{2\beta_2},$$

where ι is the summation vector. The conditional pseudo-true value β_α^* maximizes the RHS of (5.39) with respect to β, and is given by

$$(5.40) \qquad \beta_\alpha^* = \begin{pmatrix} (\Xi_G'\Xi_G)^{-1}\Xi_G'\mu_1 \\ \frac{1}{T}\iota'\mu_2 + \frac{1}{T}\mu_1'\left[I - \Xi_G(\Xi_G'\Xi_G)^{-1}\Xi_G'\right]\mu_1 \end{pmatrix},$$

provided that Ξ_G has full column rank. The average distance between F_α and $\mathcal{G}$ follows upon subtracting (5.39), with $\beta = \beta_\alpha^*$, from (5.38),

$$(5.41) \qquad \bar{I}(F_\alpha, \mathcal{G}) = \frac{1}{2}\log\left(\frac{\beta_{2\alpha}^*}{\alpha_2}\right) - \frac{1}{T}\iota'\Xi_F\alpha_1$$

where $\beta_{2\alpha}^*$ is the conditional variance of $G_{\beta_\alpha^*}$.

As in Example 5.3, the conditional mean and variance of $G_{\beta_\alpha^*}$ have obvious interpretations in terms of orthogonal projections in R^T. $\qquad\square$

6. Conclusion

Using the KLIC as a directional distance measure between distributions, it is possible, under suitable regularity conditions, to map the distributions of a given parametric family into another parametric family, such that the KLIC is minimal for each pair of the map. This approach formalizes the idea of approximating the distributions of a given family by those of another given family. The overall quality of the approximation is quantified by the resulting value of the KLIC. The approximating distribution is called the pseudo-true distribution. For any given parametrizations of the families involved, the map may also be described in terms of the pseudo-true value, which maps the parameter space of the first family into the parameter space of the second family. The whole approach bears a strong relationship with pseudo-ML estimation, since the pseudo-true value is the almost sure limit of the pseudo-ML estimator for a suitably chosen parametrization. Since the definition of encompassing (see Chapter 2) is based on the approximations induced by the KLIC, the development of a theory of statistical inference relative to encompassing can greatly benefit from the results established by the pseudo-ML theory, as we will show in Chapter 3.

As we have demonstrated, the pseudo-true value does not always exist. In such a case, the pseudo-ML estimator does not converge almost surely, regardless of the parametrization chosen. This simply means that the KLIC is unbounded or has no unique minimum, so that we cannot define a 'best' approximation according to the KLIC. As of yet, it remains an open question whether the adoption of another distance measure between distributions can resolve this problem, while at the same time preserving analytical tractability in applications. Still more difficult seems the development of a general theory of inference relative to such approximations, since then there seems to be no scope for borrowing the results from the pseudo-ML theory.

Chapter 2

Encompassing

1. Introduction

In many scientific fields, it is a widespread idea that a theory should be able to explain the results obtained by other, competing theories. This idea, although implicitly adhered to by many authors, has only recently been introduced formally for the purpose of evaluating empirical statistical models. In the econometrics literature, it became known as the encompassing principle through the work of Hendry, Mizon and Richard, who advocated and formalized it in a number of papers (Hendry–Richard [1982, 1983, 1990], Mizon [1984] and Mizon–Richard [1986]). See also Govaerts [1987], Smith [1993] and White [1994]. A recent comprehensive treatment of encompassing is given by Gouriéroux–Monfort [1995]. Loosely speaking, a model $\mathcal{F}$ is said to encompass a model $\mathcal{G}$ if it can account for the results obtained within the context of $\mathcal{G}$.

Historically, the theory of encompassing has mainly been developed as a theory of hypotheses testing, particularly non-nested hypotheses testing. In this spirit, the early definitions of encompassing were stated in terms of random quantities. A more precise definition involves the comparison of a pair of pseudo-true values which are defined by the true data generating process (DGP) and the two parametric models under study—see Hendry–Richard [1990] and Gouriéroux–Monfort [1992]. This definition is general enough to be applicable to any pair of models which assume the same dependent and conditioning variables, in a sense to be made clear below. Hence, the models $\mathcal{F}$ and $\mathcal{G}$ may be nested, disjoint or overlapping, and both, only one or neither may be correctly specified.

In this chapter, building on the work of the above mentioned authors, encompassing is defined as a binary relation on a given class $\mathcal{M}$ of parametric families of distributions. The relation is based on the approximations of some part of the DGP by the elements of $\mathcal{M}$. Using the approximations as

31

induced by the KLIC, $\mathcal{F}$ is said to encompass $\mathcal{G}$ if the pseudo-true distribution of $\mathcal{G}$ relative to the DGP is equal to the pseudo-true distribution of $\mathcal{G}$ relative to the pseudo-true distribution of $\mathcal{F}$ relative to the DGP. An equivalent formulation is in terms of a pair of pseudo-true parameter values. We study the properties of the encompassing relation and discuss its usefulness for empirical model building. Throughout, we assume an independent and identical joint DGP for the dependent and the conditioning variables.

The KLIC plays a key role in the definition of the pseudo-true values, and hence in the definition of encompassing. On the other hand, the KLIC induces a preordering on $\mathcal{M}$ which can be used as a model selection criterion (see, e.g., Akaike [1973]). Specifically, $\mathcal{F}$ is then preferred to $\mathcal{G}$ if the distance between the DGP and $\mathcal{F}$, as measured by the KLIC, is smaller than the distance between the DGP and $\mathcal{G}$. The question arises then of comparing the encompassing relation with the preordering defined by the KLIC. From a formal point of view we have little to say about this, although we show below that the encompassing relation is not transitive and hence does not induce a preordering on $\mathcal{M}$. From a less formal point of view, it appears that the model choice approach is of a more descriptive nature and is better suited if one is interested in maximizing the predictive power of the model. On the other hand, the encompassing approach is more interested in the parameters of the models and hence, provided that these parameters have an interesting interpretation *outside* the models, is more structural in nature.

This chapter is organized as follows. In Section 2, we discuss at some length the basic framework and terminology from the point of view of an empirical model builder who wants to compare two different parametric empirical models. A formal definition of the encompassing relation is given in Section 3. We study the theoretical properties of this relation in Section 4, and draw some conclusions regarding modelling strategy in Section 5. In Section 6, we reconsider the examples given in Chapter 1 and study encompassing relations in the context of these examples. Section 7 summarizes and concludes this chapter.

2. The data generating process and empirical models

Empirical models all try to capture relevant features of some part of reality. In order to compare them and analyze their relative merits, it is convenient to postulate the existence of a process which actually generates the data. Although unknown and unlikely ever to be discovered for a complex reality such as an economy, the data generating process (DGP) provides a basic framework for reasoning and interpretation. We conceive the DGP as *the* distribution of a random vector W, which is the full set of relevant and observable real world phenomena. It has to be noted that some authors, in particular those who introduced this concept into the literature, define the DGP as a *set* of distributions (or, equivalently, as a *set* of density functions) indexed by a finite dimensional parameter vector—see, e.g., Richard [1980], Hendry–Richard [1982, 1983], Engle–Hendry–Richard [1983], Hendry–Pagan–Sargan [1983] and Mizon–Richard [1986]. We feel that each element of such a set defines a different process for generating data, and that the process which actually generates the real world data must be unique. Hence, we adhere to the definition of the DGP as the (unknown) distribution which corresponds to the process generating the real world data.

Since, conceptually, reality is conflated with the DGP, the number of variables in W is likely to be very large and far beyond the reach of an empirical model builder, who usually concentrates on a subset of variables only. This may be interpreted as an implicit marginalization with respect to the variables Z not considered. Clearly, the final goal of the researcher should govern the choice of the retained variables, as well as their further partitioning into a set of 'endogenous' variables Y, which are considered to be of primary interest, and a set of 'exogenous' or conditioning variables X. It is important to realize that, if Y and X are disjoint but otherwise arbitrary subsets of W, the conditional distribution of Y, given X, derives unambiguously from the DGP, by marginalizing with respect to the variables in W, but not in Y or X, and subsequent conditioning on X. If we partition W into $W = (X, Y, Z)$ and denote the distribution function of a random vector by $\mathrm{DF}(\cdot)$, the distribution function associated with the DGP can be written as

$$(2.1) \qquad \mathrm{DF}(W) = \mathrm{DF}(Z|X, Y)\, \mathrm{DF}(Y|X)\, \mathrm{DF}(X).$$

At this point, the distribution of Y, given X, is defined to be the distribution of interest to the researcher. It is assumed that neither the conditional distribution of Z, given X and Y, nor the marginal distribution of X is of direct interest.

Marginalizing and conditioning are referred to as reductions. All reductions are unique. The reverse step, i.e., specifying a joint distribution from which a given marginal or conditional distribution derives, is called an extension. Extensions are clearly non-unique. From this perspective, the researcher's choices of Y and X have unambiguous interpretations in terms of well-defined reductions of the DGP. The marginalization and the conditioning steps are, theoretically speaking, harmless, although any marginalization or conditioning with respect to whatever variable obviously decreases the substantive scope of the modelling activity.

It is convenient to introduce some notation. Denote the conditional distribution of Y, given X, by D, and the marginal distribution of X by P. Hence, the distribution functions associated with D and P are $DF(Y|X)$ and $DF(X)$, respectively. The conditional distribution of Z, given X and Y, will play little further role. Let the random vectors Y and X have dimension k and l, respectively, i.e., Y and X take values in the Euclidean space R^k and R^l, respectively. Furthermore, let $\mathcal{M}$ be the class of all parametric families of k-dimensional conditional distributions, given X.

Given the choice of Y and X, a *parametric model* is then defined as a parametric family $\mathcal{F} = \{F_\alpha | \alpha \in \Omega_{\mathcal{F}} \subset R^m\} \in \mathcal{M}$, for some non-negative integer m. Loosely speaking, $\mathcal{M}$ is the set of all parametric models of Y, given X. In many circumstances, only a subset S of the conditioning variables X actually appear in the distributions of $\mathcal{F}$. The deliberate omission of conditioning variables is tantamount to saying that the distributions in $\mathcal{F}$ only depend on X through S, which is a conditional independence assumption. In specifying $\mathcal{F}$, it is hoped that particular choices of the parameter α would yield useful approximations to D.

We argued that the choice of endogenous variables Y and conditioning variables X should be motivated by the particular purpose of the model, so that no specification error whatsoever results from this choice. Specification errors may and are likely to enter in the form of wrong conditional independence assumptions or, more generally, incorrect functional specification of the distributions in $\mathcal{F}$. By definition, a model $\mathcal{F}$ is correctly specified if

34

$D \in \mathcal{F}$. Otherwise, it is misspecified.

Consider now a second parametric model $\mathcal{G} = \{G_\beta | \beta \in \Omega_\mathcal{G} \subset R^n\} \in \mathcal{M}$, which is seen as a competitor to $\mathcal{F}$. It is useful to think of $\mathcal{F}$ and $\mathcal{G}$ as resulting from competing theories about the phenomena involved. The task of the researcher is then to judge and compare their qualities as approximations to D. The explicit recognition that the 'Axiom of Correct Specification' (Leamer [1978]) is unlikely to hold for empirical models calls for a framework in which to interpret and compare possibly misspecified models. It seems only natural, if not trivial, that the interpretation of the models under scrutiny takes place within the context of the underlying DGP.

It has to be emphasized that a formal comparison of any two models $\mathcal{F}$ and $\mathcal{G}$ requires them to be defined with respect to the distribution of the same variables. More precisely, two models can only be fully compared if they assume the same sets of endogenous variables Y and conditioning variables X. If this is not the case, the interpretation of $\mathcal{F}$ and $\mathcal{G}$ as seen from the DGP remains unambiguous, but $\mathcal{F}$ and $\mathcal{G}$ cannot be compared *sensu stricto*, as they embody statements pertaining to different aspects of reality. The distributions of one or both of the models may then have to be modified by reductions or extensions, so as to provide a common distributional framework in which to compare the modified models—see also Mizon [1984] for a discussion of this issue. No general rules for such modifications are available, however, and besides being non-unique, they may be in conflict with the original spirit of the models. Some examples may help to clarify this point. Without ambiguity, we can consider any random vector as a *set* of scalar random variables, enabling us to use the set operations $\cap$, $\cup$ and $\setminus$.

Example 2.1. Let $\mathcal{F}$ be a model of Y, given X, and $\mathcal{G}$ a model of Y^*, given X, such that Y and Y^* partially overlap. Then, by marginalizing, one may reduce the distributions of both models to conditional distributions of $Y \cap Y^*$, given X, or one may extend them to conditional distributions of $Y \cup Y^*$, given X, such that the distributions of $\mathcal{F}$ and $\mathcal{G}$ derive from their respective extended versions. Reductions and extensions may also be combined in numerous ways. □

In the following example, taken from Gouriéroux–Monfort [1989], we indicate how a market equilibrium model can be compared with a fixed price disequilibrium model.

35

Example 2.2. Consider a market where, by a temporary change of notation, D, S, Q, P and X denote demand, supply, transaction volume, price and other market determinants, respectively. It is assumed that Q, P and X are observable, while D and S need not be so. Let both the equilibrium model and the disequilibrium model specify the following demand and supply equations:

$$(2.2) \qquad\qquad D = \gamma_1 P + \delta'_1 X + u_1,$$

$$(2.3) \qquad\qquad S = \gamma_2 P + \delta'_2 X + u_2,$$

where $(\gamma_1, \gamma_2, \delta'_1, \delta'_2)'$ is a parameter vector, and the distribution of $(u_1, u_2)'$ lies within a specified parametric family. Furthermore, the equilibrium model specifies

$$(2.4) \qquad\qquad Q = D = S,$$

and considers Q and P as endogenous and X as exogenous. The conditional distribution of Q and P, given X, is determined by (2.2)–(2.4). On the other hand, the fixed price disequilibrium model specifies

$$(2.5) \qquad\qquad Q = \min\{D, S\},$$

and considers Q as endogenous and P and X as exogenous. In this model, the conditional distribution of Q, given P and X, is determined by (2.2)–(2.3) and (2.5). A reduction of the equilibrium model by conditioning on P provides a common distributional framework in which both models can be compared. $\qquad\qquad\qquad\qquad\qquad\qquad\qquad\qquad\qquad\qquad\qquad\qquad\square$

Since extensions are non-unique and highly arbitrary, it may be advisable to apply only reductions to modify the distributions of the models. Furthermore, considering the Kullback-Leibler distance between the true distribution of a set of variables and a model thereof, it is easily seen that an extension of the distributions of the model adds a non-negative term to the distance. This term is zero if and only if the extended part is correctly specified. Otherwise said, an extension can only render the model 'more misspecified'. The opposite holds for a reduction. Unfortunately, it is not always possible to apply only reductions, as the following example demonstrates.

Example 2.3. Let $\mathcal{F}$ be a model of Y, given X, and $\mathcal{G}$ a model of Y, given X^*, such that X is a proper subset of X^*. Then, reductions of the distributions of $\mathcal{G}$ always have the full set X^* among the conditioning variables, whereas reductions of the distributions of $\mathcal{F}$ never have, since the variables $X^* \setminus X$ do not appear in the distributions of $\mathcal{F}$. There are many ways to proceed then, although none of them is satisfactory. One may either extend the distributions of $\mathcal{F}$ to conditional distributions of $Y \cup (X^* \setminus X)$, given X, and subsequently reduce them by conditioning with respect to $X^* \setminus X$, or extend the distributions of $\mathcal{G}$ to conditional distributions of $Y \cup (X^* \setminus X)$, given X, and subsequently reduce them by marginalizing with respect to $X^* \setminus X$, or even extend the distributions of both models to conditional distributions of $Y \cup (X^* \setminus X)$, given X. Many intermediate forms are also possible. □

In the sequel, we assume that $(\mathcal{F}, \mathcal{G}) \in \mathcal{M} \times \mathcal{M}$, i.e., $\mathcal{F}$ and $\mathcal{G}$ both have Y as endogenous variables and X as conditioning variables. We will define encompassing as a binary relation on $\mathcal{M}$, relative to D and P. That is, for any pair $(\mathcal{F}, \mathcal{G}) \in \mathcal{M} \times \mathcal{M}$, we will define whether $\mathcal{F}$ encompasses $\mathcal{G}$ or not, relative to D and P. From a formal point of view, an encompassing relation can be defined relative to any k-dimensional conditional distribution and any l-dimensional marginal distribution. However, the relevance of the encompassing principle for empirical model building becomes clear by considering D and P as the true distributions of the real world phenomena $Y|X$ and X. The definition of the encompassing relation relies on the comparison of a pair of pseudo-true values which are defined in the next section.

3. Definition of encompassing

As in Chapter 1, we consider conditional distributions on (A, σ_A), given X. A is the Euclidean space R^k, σ_A is the Borel σ-field on A, and X is a random vector defined on a probability space (C, σ_C, P), where $C = R^l$ and σ_C is the completion of the Borel σ-field on C with respect to the probability measure P, the distribution of X. Let $\mathcal{D}$ be the set of all conditional distributions on (A, σ_A), given X, and let $\mathcal{M}$ be the class of all parametric families of such distributions. Note that $\mathcal{D} \notin \mathcal{M}$ since $\mathcal{D}$ can not be mapped one-to-one into a Euclidean space, and that $\mathcal{D}$ is the union of all $\mathcal{F} \in \mathcal{M}$.

Let $\mathcal{F} = \{F_\alpha | \alpha \in \Omega_\mathcal{F}\}$ and $\mathcal{G} = \{G_\beta | \beta \in \Omega_\mathcal{G}\}$ be elements of $\mathcal{M}$. For each $D \in \mathcal{D}$, the pseudo-true values of α and β relative to D and P are denoted by α_D and β_D, respectively, provided that these pseudo-true values exist. (Here and in the sequel, the existence of a pseudo-true value is to be understood in the stronger sense of existence and uniqueness.) For each $(D, \mathcal{F}) \in \mathcal{D} \times \mathcal{M}$ such that α_D exists, the pseudo-true value of β relative to F_{α_D} and P is denoted by β_{α_D}, provided that this pseudo-true value exists.

Following Hendry–Richard [1990] and Gouriéroux–Monfort [1995], we define for a given P and for each $D \in \mathcal{D}$ an encompassing relation, denoted $\mathcal{E}_D^P$, as a binary relation on $\mathcal{M}$ which involves the comparison of β_D and β_{α_D}.

Definition 3.1. *For all $(\mathcal{F}, \mathcal{G}) \in \mathcal{M} \times \mathcal{M}$, we say that $\mathcal{F}$ provides a complete parametric encompassing of $\mathcal{G}$ relative to D and P, written $\mathcal{F} \; \mathcal{E}_D^P \; \mathcal{G}$, if and only if*

$$(3.1) \qquad\qquad \beta_D = \beta_{\alpha_D}.$$

Here and in the sequel, the equality is to be interpreted in the sense that, if either side exists so does the other, and the two are equal. If no conditioning variables are present, the encompassing relation is written $\mathcal{E}_D$. Notice that $\beta_D = \beta_{\alpha_D}$ is equivalent to $G_{\beta_D} = G_{\beta_{\alpha_D}}$. When $\mathcal{F} \; \mathcal{E}_D^P \; \mathcal{G}$, we also say that $\mathcal{F}$ encompasses $\mathcal{G}$ relative to D and P. If furthermore the reference to D and P is clear from the context, we will not mention it explicitly. The possible non-existence of the pseudo-true values makes the definition of $\mathcal{E}_D^P$ look somewhat complicated. The definition is ultimately justified by the properties which result from it. In order to have $\mathcal{F} \; \mathcal{E}_D^P \; \mathcal{G}$, β_D must exist if and only if β_{α_D} exists, and the two must be equal upon existence. Thus, if β_D exists, then α_D and β_{α_D} must both exist. Also, if β_D does not exist and α_D exists, then β_{α_D} must not exist. If $\beta_D \neq \beta_{\alpha_D}$, we write $\mathcal{F} \; \not{\mathcal{E}}_D^P \; \mathcal{G}$. We also adopt the following terminology from Hendry–Richard [1990] and Gouriéroux–Monfort [1995]. If only a subvector of $\beta_D - \beta_{\alpha_D}$ is zero, we say that $\mathcal{F}$ provides an *incomplete parametric encompassing* of $\mathcal{G}$. If $\mathcal{F}$ is a subset of $\mathcal{G}$ and $\mathcal{F} \; \mathcal{E}_D^P \; \mathcal{G}$, $\mathcal{F}$ is said to *parsimoniously encompass* $\mathcal{G}$. Finally, if $\mathcal{F} \; \mathcal{E}_D^P \; \mathcal{G}$ and $\mathcal{G} \; \mathcal{E}_D^P \; \mathcal{F}$, we say that $\mathcal{F}$ and $\mathcal{G}$ *mutually encompass each other*.

$\mathcal{F} \, \mathcal{E}_D^P \, \mathcal{G}$ embodies the property that the closest approximation to D by $\mathcal{G}$ coincides with the closest approximation by $\mathcal{G}$ to the closest approximation to D by $\mathcal{F}$. If $\mathcal{F} \, \mathcal{E}_D^P \, \mathcal{G}$, then these approximations do not coincide or one and only one of the two exists. If we interpret parametric families of distributions by reference to their pseudo-true values, then $\mathcal{F} \, \mathcal{E}_D^P \, \mathcal{G}$ if and only if the interpretation of $\mathcal{G}$, seen from D, coincides with the interpretation of $\mathcal{G}$, seen from the interpretation of $\mathcal{F}$, seen from D. Thus, encompassing is a formalization of the idea that, looking at $\mathcal{G}$ from D, one observes the same thing as looking at $\mathcal{G}$ from D through $\mathcal{F}$. Figure 1 gives a graphical illustration of the encompassing relations relative to P and to the distributions D and E for disjoint families $\mathcal{F}$ and $\mathcal{G}$.

Figure 1. *Encompassing relations:* $\mathcal{F} \, \mathcal{E}_D^P \, \mathcal{G}$ *and* $\mathcal{F} \, \mathcal{E}_E^P \, \mathcal{G}$.

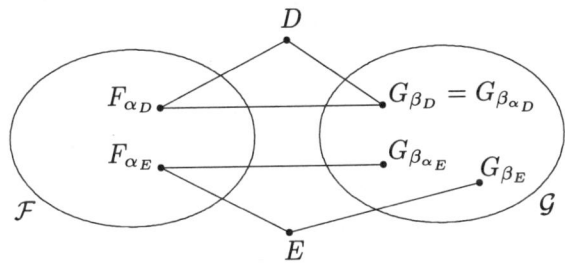

Provided that β_D and β_{α_D} exist, the average difference between the log-densities of G_{β_D} and $G_{\beta_{\alpha_D}}$ can be interpreted as a measure of non-encompassing, say $\mathcal{NE}_D^P$. In the notation of Chapter 1,

$$(3.2) \quad \mathcal{NE}_D^P \, (\mathcal{F}, \mathcal{G}) = E_P E_D [\log g(Y|X; \beta_D) - \log g(Y|X; \beta_{\alpha_D})]$$
$$= I_P(D, G_{\beta_{\alpha_D}}) - I_P(D, G_{\beta_D}) \geq 0,$$

showing that $\mathcal{NE}_D^P$ has a nice interpretation in terms of the KLIC. Note also that $\mathcal{NE}_D^P$ is invariant with respect to the parametrizations of $\mathcal{F}$ and $\mathcal{G}$.

Given a vector $(x_1, \ldots, x_T)$ of independent realizations of X, one may also define an encompassing relation by referring to $(x_1, \ldots, x_T)$ instead of P. The definition of conditional pseudo-true values induces this definition. Upon existence, denote the conditional pseudo-true values of α and β relative to D, given $(x_1, \ldots, x_T)$, by α_D^* and β_D^*, respectively, and the conditional pseudo-true value of β relative to $F_{\alpha_D^*}$, given $(x_1, \ldots, x_T)$, by $\beta_{\alpha_D^*}^*$.

Definition 3.2. For all $(\mathcal{F}, \mathcal{G}) \in \mathcal{M} \times \mathcal{M}$ and for a given $(x_1, \ldots, x_T)$, we say that $\mathcal{F}$ provides a conditional complete parametric encompassing of $\mathcal{G}$ relative to D, given $(x_1, \ldots, x_T)$, written $\mathcal{F} \; \mathcal{E}_D^* \; \mathcal{G}$, if and only if

$$(3.3) \qquad \beta_D^* = \beta_{\alpha_D^*}^*.$$

A conditional measure of non-encompassing may be defined as

$$(3.4) \qquad \mathcal{N}\mathcal{E}_D^* \, (\mathcal{F}, \mathcal{G}) = \frac{1}{T} \sum_{t=1}^{T} E_D[\log g(Y|x_t; \beta_D^*) - \log g(Y|x_t; \beta_{\alpha_D^*}^*)]$$

$$= \bar{I}\left(D, G_{\beta_{\alpha_D^*}^*}\right) - \bar{I}\left(D, G_{\beta_D^*}\right).$$

Note that, if the unconditional and conditional pseudo-true values (i.e., α_D and α_D^*, etc.) are asymptotically equivalent as $T \to \infty$, then

$$(3.5) \qquad \mathcal{F} \; \mathcal{E}_D^P \; \mathcal{G} \iff \beta_D^* - \beta_{\alpha_D^*}^* \xrightarrow[P]{a.s.} 0,$$

although this does not imply the asymptotic equivalence of $\mathcal{F} \; \mathcal{E}_D^P \; \mathcal{G}$ and $\mathcal{F} \; \mathcal{E}_D^* \; \mathcal{G}$ sensu stricto. The unconditional and conditional measures of non-encompassing, however, are asymptotically equivalent:

$$(3.6) \qquad \mathcal{N}\mathcal{E}_D^* \, (\mathcal{F}, \mathcal{G}) \xrightarrow[P]{a.s.} \mathcal{N}\mathcal{E}_D^P \, (\mathcal{F}, \mathcal{G}).$$

4. Properties of the encompassing relation

In this section, we present some properties of the encompassing relation $\mathcal{E}_D^P$ which we think have relevant implications regarding modelling strategy, to be discussed in Section 5 below. Properties 4.2, 4.5, 4.10, 4.16 and, to some extent, Properties 4.8, 4.11, 4.14 and 4.15 have already been given by Gouriéroux–Monfort [1995]. We study $\mathcal{E}_D^P$ in the general framework of conditional distributions, including unconditional distributions as a special case. Thus, D is a conditional distribution on (A, σ_A), given some random vector X with distribution P, and $\mathcal{M}$ is the class of parametric families of conditional distributions on (A, σ_A), given X. All the properties of $\mathcal{E}_D^P$ carry over to $\mathcal{E}_D^*$.

Throughout, $\mathcal{F}, \mathcal{G}, \mathcal{P}$ and $\mathcal{Q}$ denote any elements of $\mathcal{M}$, and are indexed by the parameter vectors α, β, γ and δ, respectively. For every $(\mathcal{F}, \mathcal{G}) \in$

$\mathcal{M} \times \mathcal{M}$, we define the following sets. The terminology is due to Gouriéroux–Monfort [1995]. The *image of* $\mathcal{F}$ *in* $\mathcal{G}$ is defined as

$$(4.1) \qquad \mathcal{G}_{\mathcal{F}} = \{G_\beta \in \mathcal{G} \mid \exists\, F_\alpha \in \mathcal{F} : \beta = \beta_\alpha\}.$$

The *reflecting set of* $\mathcal{G}$ *from* $\mathcal{F}$ is defined as the fixed point set

$$(4.2) \qquad \mathcal{G}_{\mathcal{F}}^R = \{G_\beta \in \mathcal{G} \mid \exists\, F_\alpha \in \mathcal{F} : \beta = \beta_\alpha \quad \text{and} \quad \alpha = \alpha_\beta\}.$$

Clearly, $\mathcal{G}_{\mathcal{F}}^R \subset \mathcal{G}_{\mathcal{F}} \subset \mathcal{G}$. Also, $\mathcal{G}_{\mathcal{F}}^R \cap \mathcal{F}_{\mathcal{G}}^R = \mathcal{G} \cap \mathcal{F}$ and, if $\mathcal{F} \subset \mathcal{G}$, then $\mathcal{F}_{\mathcal{G}}^R = \mathcal{G}_{\mathcal{F}}^R = \mathcal{F}_{\mathcal{G}} = \mathcal{G}_{\mathcal{F}} = \mathcal{F}$. The reflecting sets may be empty when $\mathcal{F} \cap \mathcal{G} = \emptyset$.

The properties given hereafter and their proofs are easily understood by means of figures like Figure 1. They are presented without discussion. The next section gives interpretations and discusses their usefulness for empirical model building.

Property 4.1. $\{D\} \ \mathcal{E}_D^P \ \mathcal{G}$.

Proof. The result follows immediately from the definition of $\mathcal{E}_D^P$. $\qquad\qquad \square$

The following property is a generalization of Property 4.1.

Property 4.2. If $D \in \mathcal{F}$, then $\mathcal{F} \ \mathcal{E}_D^P \ \mathcal{G}$.

Proof. If $D \in \mathcal{F}$, then $D = F_{\alpha_D}$. Thus, $\beta_D = \beta_{\alpha_D}$. $\qquad\qquad \square$

Property 4.3. If $D \in \mathcal{G}$, then

$$(4.3) \qquad\qquad \mathcal{F} \ \mathcal{E}_D^P \ \mathcal{G} \iff D \in \mathcal{G}_{\mathcal{F}}^R.$$

Proof. The result follows immediately from the definitions of $\mathcal{E}_D^P$ and $\mathcal{G}_{\mathcal{F}}^R$. $\qquad\qquad \square$

For nested families of distributions, (i.e., where one family is a subset of the other), we have the following properties.

Property 4.4. If $D \notin \mathcal{F}$, then

$$(4.4) \qquad\qquad \forall \mathcal{G} : \mathcal{F} \subset \mathcal{G} \quad \text{and} \quad D \in \mathcal{G} \Rightarrow \mathcal{F} \ \mathcal{E}_D^P \ \mathcal{G}.$$

Proof. Let $D \notin \mathcal{F} \subset \mathcal{G}$ and $D \in \mathcal{G}$. Then, $D = G_{\beta_D}$ and $F_{\alpha_D} = G_{\beta_{\alpha_D}} \neq D$, wherefrom $G_{\beta_D} \neq G_{\beta_{\alpha_D}}$. $\qquad\qquad \square$

41

Property 4.5. If $\mathcal{F} \subset \mathcal{G}$, then

$$(4.5) \qquad \mathcal{F} \; \mathcal{E}_D^P \; \mathcal{G} \iff F_{\alpha_D} = G_{\beta_D},$$

$$(4.6) \qquad \qquad \iff I_P(D, \mathcal{F}) = I_P(D, \mathcal{G}).$$

Proof. If $\mathcal{F} \subset \mathcal{G}$, then $F_{\alpha_D} = G_{\beta_{\alpha_D}}$. Therefore, $F_{\alpha_D} = G_{\beta_D}$ is equivalent to $G_{\beta_{\alpha_D}} = G_{\beta_D}$, i.e., to $\mathcal{F} \; \mathcal{E}_D^P \; \mathcal{G}$. Also, $I_P(D, \mathcal{F}) = I_P(D, F_{\alpha_D}) = I_P(D, G_{\beta_{\alpha_D}})$, which in turn is equal to $I_P(D, \mathcal{G}) = I_P(D, G_{\beta_D})$ iff $\mathcal{F} \; \mathcal{E}_D^P \; \mathcal{G}$. $\qquad \square$

Property 4.6. If β_D exists and $G_{\beta_D} \in \mathcal{F} \subset \mathcal{G}$, then $\mathcal{F} \; \mathcal{E}_D^P \; \mathcal{G}$.

Proof. If β_D exists and $G_{\beta_D} \in \mathcal{F} \subset \mathcal{G}$, then α_D exists, and $G_{\beta_D} = F_{\alpha_D} = G_{\beta_{\alpha_D}}$. $\qquad \square$

Property 4.7. If γ_D exists and $P_{\gamma_D} \in \mathcal{F} \subset P$ and $P_{\gamma_D} \in \mathcal{G} \subset P$, then $\mathcal{F} \; \mathcal{E}_D^P \; \mathcal{G}$.

Proof. If γ_D exists, $P_{\gamma_D} \in \mathcal{F} \subset P$ and $P_{\gamma_D} \in \mathcal{G} \subset P$, then α_D and β_D both exist, and $F_{\alpha_D} = P_{\gamma_D} = G_{\beta_D}$, implying $G_{\beta_{\alpha_D}} = F_{\alpha_D} = G_{\beta_D}$. $\qquad \square$

Property 4.8. If β_D exists and $\mathcal{F} \subset \mathcal{G}$, then
$$(4.7)$$
$$\mathcal{F} \; \mathcal{E}_D^P \; \mathcal{G} \iff (\forall \, P, Q : \mathcal{F} \subset P \subset \mathcal{G} \quad \text{and} \quad \mathcal{F} \subset Q \subset \mathcal{G} \Rightarrow P \; \mathcal{E}_D^P \; Q).$$

Proof. Let β_D exist and let $\mathcal{F} \subset P \subset \mathcal{G}$ and $\mathcal{F} \subset Q \subset \mathcal{G}$. If $\mathcal{F} \; \mathcal{E}_D^P \; \mathcal{G}$, then α_D exists and it follows from (4.5) that $G_{\beta_D} \in P$ and $G_{\beta_D} \in Q$. Hence, from Property 4.7, $P \; \mathcal{E}_D^P \; Q$. The reverse implication in (4.7) follows immediately by taking $P = \mathcal{F}$ and $Q = \mathcal{G}$. $\qquad \square$

Property 4.9. If β_D exists and $\mathcal{F} \subset \mathcal{G}$, then

$$(4.8) \qquad \mathcal{F} \; \mathcal{E}_D^P \; \mathcal{G} \iff (\forall P : \mathcal{F} \; \mathcal{E}_D^P \; P \iff \mathcal{G} \; \mathcal{E}_D^P \; P).$$

Proof. If β_D exists, $\mathcal{F} \subset \mathcal{G}$ and $\mathcal{F} \; \mathcal{E}_D^P \; \mathcal{G}$, then α_D and β_D both exist and, by (4.5), $F_{\alpha_D} = G_{\beta_D}$. Thus, $\gamma_{\alpha_D} = \gamma_{\beta_D}$. Consequently, $\mathcal{F} \; \mathcal{E}_D^P \; P \iff \mathcal{G} \; \mathcal{E}_D^P \; P$. To prove the reverse implication, let $\mathcal{F} \subset \mathcal{G}$ and let β_D exist. Then, if $\mathcal{F} \; \cancel{\mathcal{E}}_D^P \; \mathcal{G}$, take $P = \mathcal{G}$, which gives $\mathcal{G} \; \mathcal{E}_D^P \; \mathcal{G}$, to see that the equivalence $\mathcal{F} \; \mathcal{E}_D^P \; P \iff \mathcal{G} \; \mathcal{E}_D^P \; P$ does not hold for all P. $\qquad \square$

When the families $\mathcal{F}$ and $\mathcal{G}$ are not necessarily nested, we have the following properties.

Property 4.10. *If β_D exists, then*

$$(4.9) \qquad \mathcal{F} \; \mathcal{E}_D^P \; \mathcal{G} \; \Rightarrow \; I_P(D, \mathcal{G}_{\mathcal{F}}) = I_P(D, \mathcal{G}).$$

Proof. If β_D exists and $\mathcal{F} \; \mathcal{E}_D^P \; \mathcal{G}$, then $G_{\beta_D} = G_{\beta_{\alpha_D}} \in \mathcal{G}_{\mathcal{F}}$. Then, from Property 4.6, it follows that $\mathcal{G}_{\mathcal{F}} \; \mathcal{E}_D^P \; \mathcal{G}$, and from (4.6) that $I_P(D, \mathcal{G}_{\mathcal{F}}) = I_P(D, \mathcal{G})$. $\square$

Property 4.11. *If β_D exists, then*
(4.10)
$$\mathcal{F} \; \mathcal{E}_D^P \; \mathcal{G} \; \Rightarrow \; (\forall \, \mathcal{P}, \mathcal{Q} : \mathcal{G}_{\mathcal{F}} \subset \mathcal{P} \subset \mathcal{G} \;\; \text{and} \;\; \mathcal{G}_{\mathcal{F}} \subset \mathcal{Q} \subset \mathcal{G} \; \Rightarrow \; \mathcal{P} \; \mathcal{E}_D^P \; \mathcal{Q}).$$

Proof. From Property 4.10 and Property 4.5, it follows that if β_D exists and $\mathcal{F} \; \mathcal{E}_D^P \; \mathcal{G}$, then $\mathcal{G}_{\mathcal{F}} \; \mathcal{E}_D^P \; \mathcal{G}$. Then, the result obtains immediately from Property 4.8. $\square$

Property 4.12. *If γ_D and δ_D both exist, $P_{\gamma_D} \in \mathcal{F} \subset \mathcal{P}$, $Q_{\delta_D} \in \mathcal{G} \subset \mathcal{Q}$ and $\mathcal{P} \; \mathcal{E}_D^P \; \mathcal{Q}$, then $\mathcal{F} \; \mathcal{E}_D^P \; \mathcal{G}$.*

Proof. If γ_D exists and $P_{\gamma_D} \in \mathcal{F} \subset \mathcal{P}$, $F_{\alpha_D} = P_{\gamma_D}$. Likewise, if δ_D exists and $Q_{\delta_D} \in \mathcal{G} \subset \mathcal{Q}$, then $G_{\beta_D} = Q_{\delta_D}$. If furthermore $\mathcal{P} \; \mathcal{E}_D^P \; \mathcal{Q}$, then $Q_{\delta_D} = Q_{\delta\gamma_D} = Q_{\delta\alpha_D} \in \mathcal{G}$, wherefrom $G_{\beta_D} = G_{\beta_{\alpha_D}}$. $\square$

Property 4.13. *If β_D exists, then*

$$(4.11) \qquad \mathcal{F} \; \mathcal{E}_D^P \; \mathcal{G} \; \Longleftrightarrow \; (\forall \, \mathcal{P} : \mathcal{G}_{\mathcal{F}} \subset \mathcal{P} \subset \mathcal{G} \; \Rightarrow \; \mathcal{F} \; \mathcal{E}_D^P \; \mathcal{P}).$$

Proof. If $\mathcal{F} \; \mathcal{E}_D^P \; \mathcal{G}$ and $\mathcal{G}_{\mathcal{F}} \subset \mathcal{P} \subset \mathcal{G}$, then $P_{\gamma_D} = G_{\beta_D} = G_{\beta_{\gamma_D}} = P_{\gamma_{\alpha_D}}$, which proves the implication. The reverse implication follows immediately by taking $\mathcal{P} = \mathcal{G}$. $\square$

With respect to mutual encompassing, we have the following properties.

Property 4.14. *If β_D exists, $\mathcal{F} \; \mathcal{E}_D^P \; \mathcal{G}$ and $\mathcal{G} \; \mathcal{E}_D^P \; \mathcal{F}$, then*
(4.12)
$$\forall \, \mathcal{P}, \mathcal{Q} : \mathcal{F}_{\mathcal{G}}^R \subset \mathcal{P} \subset \mathcal{F} \;\; \text{and} \;\; \mathcal{G}_{\mathcal{F}}^R \subset \mathcal{Q} \subset \mathcal{G} \; \Rightarrow \; \mathcal{P} \; \mathcal{E}_D^P \; \mathcal{Q} \;\; \text{and} \;\; \mathcal{Q} \; \mathcal{E}_D^P \; \mathcal{P}.$$

Proof. If β_D exists, $\mathcal{F} \; \mathcal{E}_D^P \; \mathcal{G}$ and $\mathcal{G} \; \mathcal{E}_D^P \; \mathcal{F}$, then $G_{\beta_D} = G_{\beta_{\alpha_D}}$ and $F_{\alpha_D} = F_{\alpha_{\beta_D}}$. Consequently, $F_{\alpha_D} \in \mathcal{F}_{\mathcal{G}}^R$ and $G_{\beta_D} \in \mathcal{G}_{\mathcal{F}}^R$. Applying Property 4.12, the result follows. $\square$

43

Property 4.15. If β_D exists, $\mathcal{F} \, \mathcal{E}_D^P \, \mathcal{G}$ and $\mathcal{G} \, \mathcal{E}_D^P \, \mathcal{F}$, then

$$(4.13) \qquad \forall \, \mathcal{P}, \mathcal{Q} : \mathcal{F}_\mathcal{G}^R \subset \mathcal{P} \subset \mathcal{F} \quad \text{and} \quad \mathcal{F}_\mathcal{G}^R \subset \mathcal{Q} \subset \mathcal{F} \Rightarrow \mathcal{P} \, \mathcal{E}_D^P \, \mathcal{Q},$$

and

$$(4.14) \qquad \forall \, \mathcal{P}, \mathcal{Q} : \mathcal{G}_\mathcal{F}^R \subset \mathcal{P} \subset \mathcal{G} \quad \text{and} \quad \mathcal{G}_\mathcal{F}^R \subset \mathcal{Q} \subset \mathcal{G} \Rightarrow \mathcal{P} \, \mathcal{E}_D^P \, \mathcal{Q}.$$

Proof. Noticing that $F_{\alpha_D} \in \mathcal{F}_\mathcal{G}^R$ and $G_{\beta_D} \in \mathcal{G}_\mathcal{F}^R$ when β_D exists and $\mathcal{F} \, \mathcal{E}_D^P \, \mathcal{G}$ and $\mathcal{G} \, \mathcal{E}_D^P \, \mathcal{F}$, the result follows from Property 4.6. $\qquad\square$

We also have the following negative results.

Property 4.16. $\mathcal{F} \supset \mathcal{G}$ does not imply $\mathcal{F} \, \mathcal{E}_D^P \, \mathcal{G}$.

Proof. The proof is subsumed within the proof of Property 4.17. $\qquad\square$

Property 4.17. $\mathcal{E}_D^P$ is not transitive.

Proof. We give an example where $\mathcal{F} \supset \mathcal{G} \supset \mathcal{P}$, $\mathcal{F} \, \mathcal{E}_D \, \mathcal{G}$, $\mathcal{G} \, \mathcal{E}_D \, \mathcal{P}$ and $\mathcal{F} \, \mathcal{\not{E}}_D \, \mathcal{P}$. The example is an extension of an example given in Gouriéroux–Monfort [1995] to prove Property 4.16. Denoting the bivariate normal distribution with mean μ and unit covariance matrix by N_μ, define $\mathcal{F}$, $\mathcal{G}$ and $\mathcal{P}$ as

$$(4.15) \qquad \mathcal{F} = \{ N_\alpha \, | \, \alpha \in \Omega_\mathcal{F} \subset R^2 \},$$

$$(4.16) \qquad \mathcal{G} = \{ N_\beta \, | \, \beta \in \Omega_\mathcal{G} \subset R^2 \},$$

$$(4.17) \qquad \mathcal{P} = \{ N_\gamma \, | \, \gamma \in \Omega_\mathcal{P} \subset R^2 \},$$

where

$$(4.18) \qquad \Omega_\mathcal{F} = \{ (x, y) \in R^2 \, | \, x \geq -1, \, y \geq -1 \},$$

$$(4.19) \qquad \Omega_\mathcal{G} = \{ (x, y) \in R^2 \, | \, |x| \leq 1, \, |y| \leq 1 \},$$

$$(4.20) \qquad \Omega_\mathcal{P} = \{ (x, y) \in R^2 \, | \, x^2 + y^2 \leq 1 \},$$

and let $D = N_\mu$ with $\mu = (-2, 2)'$. The point μ, the cone $\Omega_{\mathcal{F}}$, the cube $\Omega_{\mathcal{G}}$, and the unit sphere $\Omega_{\mathcal{P}}$ are depicted in Figure 2. Clearly, $D \notin \mathcal{F} \supset \mathcal{G} \supset \mathcal{P}$. The pseudo-true values are found as the solutions of problems involving the minimization of the Euclidean distance in R^2. It follows easily that $\beta_D = \beta_{\alpha_D}$, $\gamma_D = \gamma_{\beta_D}$ and $\gamma_D \neq \gamma_{\alpha_D}$, i.e., $\mathcal{F} \, \mathcal{E}_D \, \mathcal{G}$, $\mathcal{G} \, \mathcal{E}_D \, \mathcal{P}$ and $\mathcal{F} \, \not{\mathcal{E}}_D \, \mathcal{P}$.

Figure 2. The non-transitivity of the encompassing relation.

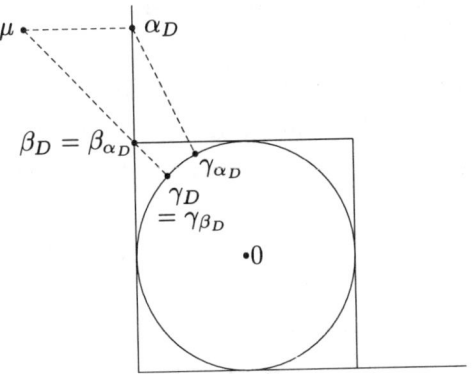

5. Encompassing and parsimony as a model reduction device

In Section 2, the random variables Y and X, the distributions D and P, and the parametric models $\mathcal{F}$ and $\mathcal{G}$ have been interpreted in the context of empirical modelling. Given any pair of competing empirical models, the question arises of assessing their relative qualities. We address this question in this section. It will be shown that the properties of the encompassing relation allow some conclusions to be drawn regarding the comparison of empirical models.

It is a fundamental property of the DGP that the performance of any model, no matter how badly misspecified, is completely determined by the DGP, whatever the definition of 'performance' may be. By this we mean that one can, at least in principle, derive from the DGP the distribution of any estimator, statistic, test outcome etc... which arises in the course of empirical model building. Also, the DGP determines the pseudo-true values associated with the models whenever they exist. This feature of the DGP is formally stated as Property 4.1.

Since, as a tool of analysis, models are intended to mimic (relevant features of) the true DGP, it is a natural question to ask whether a model can

45

mimic the DGP, in the sense that it is able to predict the performance of its competitors correctly. Note that, in the given context, the models are designed in the hope that they yield useful approximations to D, the conditional distribution of Y, given X, which constitutes a part of the DGP—see (2.1). Hence, they are to be judged according to this particular goal. An important aspect of 'performance' concerns the pseudo-true parameter value of a model, given its strong link with ML estimation. For a model to be considered as a good approximation to the DGP, it would therefore be required to encompass its rival models. A failure of $\mathcal{F}$ to encompass $\mathcal{G}$ reveals shortcomings of $\mathcal{F}$ as an approximation to the DGP, whereas the finding that $\mathcal{F}$ encompasses $\mathcal{G}$ is evidence in favour of $\mathcal{F}$. Importantly, $\mathcal{F}$ need not be correctly specified to encompass $\mathcal{G}$. However, if $\mathcal{F}$ is correctly specified, then $\mathcal{F}$ encompasses any model—see Property 4.2. Hence, encompassing is a necessary but not a sufficient condition for correct specification. Also, if $\mathcal{F}$ is misspecified, there are always models which $\mathcal{F}$ does not encompass, as is revealed by Property 4.4. This property essentially shows that any correctly specified model contains enough information to invalidate any misspecified model. In order to exploit the potential of this property, it is important to focus on theoretically plausible models which we seek to encompass by the model of primary interest.

The parsimony principle in empirical modelling is like Occam's razor: if a submodel has all the desirable properties of a larger model, we only need to consider the submodel. If we concentrate on the encompassing properties of a model, the parsimony principle allows a meaningful reduction of the number of empirical models. Such a reduction is to be distinguished from the reduction of a given distribution, which involves a marginalization or conditioning, as discussed in Section 2. Notice that, if for each $(\mathcal{F}, \mathcal{G}) \in \mathcal{M} \times \mathcal{M}$ we knew whether $\mathcal{F} \; \mathcal{E}_D^P \; \mathcal{G}$ or $\mathcal{F} \; \mathcal{E}_D^P \; \mathcal{G}$, taking the parsimony principle to the limit would lead us to consider only D, since $\{D\}$ is the smallest model which encompasses all models. However, D is unknown and complete knowledge of $\mathcal{E}_D^P$ is out of the question. It is therefore of interest to exploit partial knowledge of $\mathcal{E}_D^P$ in order to reduce the number of empirical models to be considered. The main tool to achieve such a reduction is Property 4.9, combined with Property 4.8, Property 4.11 or, in the case of mutual encompassing, Property 4.15. Notice that, since all these properties require the existence of a pseudo-true value relative to the DGP, the reduction necessarily relies on this assumption, which henceforth will not be mentioned explicitly. The starting point is Property 4.9, which states that if $\mathcal{F}$ parsimoniously encompasses $\mathcal{G}$, then both models have identical encompassing properties with respect to all other models. According to the parsimony principle and Property 4.8, we only need to consider $\mathcal{F}$, and we may forget about all other models $\mathcal{P}$ satisfying $\mathcal{F} \subset \mathcal{P} \subset \mathcal{G}$. Suppose now that $\mathcal{F}\ \mathcal{E}_D^P\ \mathcal{G}$ and that $\mathcal{F}$ and $\mathcal{G}$ are not necessarily nested. Then, from Property 4.11, $\mathcal{G}_\mathcal{F}$ parsimoniously encompasses $\mathcal{G}$. Recall that $\mathcal{G}_\mathcal{F}$ is the image of $\mathcal{F}$ in $\mathcal{G}$ and is therefore a subset of $\mathcal{G}$. Thus, we only need to consider $\mathcal{G}_\mathcal{F}$ from the set of models $\mathcal{P}$ satisfying $\mathcal{G}_\mathcal{F} \subset \mathcal{P} \subset \mathcal{G}$. The reduction is even more substantial in the case of mutual encompassing. If $\mathcal{F}$ and $\mathcal{G}$ mutually encompass each other, then, from Property 4.15, we need only to retain the reflecting sets $\mathcal{F}_\mathcal{G}^R$ and $\mathcal{G}_\mathcal{F}^R$ from the set of models P satisfying $\mathcal{F}_\mathcal{G}^R \subset \mathcal{P} \subset \mathcal{F}$ or $\mathcal{G}_\mathcal{F}^R \subset \mathcal{P} \subset \mathcal{G}$.

Some negative conclusions have to be drawn from the properties of $\mathcal{E}_D^P$ listed in Section 4. Most importantly, $\mathcal{E}_D^P$ does not induce a preordering on $\mathcal{M}$, since $\mathcal{E}_D^P$ is not transitive (Property 4.17). Perhaps $\mathcal{M}$ is too large to allow it to be meaningfully and sufficiently structured by $\mathcal{E}_D^P$. It is an open question whether the ideas of encompassing and a preordering on models can be reconciled, perhaps at the cost of altering the definition of encompassing or by considering only subclasses of $\mathcal{M}$. As of yet, the only thing we can say is that parsimonious encompassing is a preordering on the subclass of $\mathcal{M}$ for which the pseudo-true values relative to D and P exist, since this relation is reflexive and, from Property 4.9, transitive. Unfortunately, this relation only compares nested models, whereas the original idea of encompassing was to extend the comparison to non-nested models. Another open question which remains largely unanswered is the comparison of $\mathcal{E}_D^P$ to the preordering on $\mathcal{M}$ induced by the KLIC.

6. Examples

In this section, we reconsider the examples of Section 1.5 and determine the encompassing relations, restricted to the class of families involved, relative to any distribution which belongs to one of the families.

Example 6.1. In the notation of Example 1.5.1, let F, G, P and Q be the standard distributions of the families involved, i.e., with zero location and unit scale. Let D be an element of $\{F, G, P, Q\}$. Then, using the results of this example, it follows that Table 1.1 contains all the necessary information to determine $\mathcal{E}_D$, restricted to $\{\mathcal{F}, \mathcal{G}, \mathcal{P}, \mathcal{Q}\}$. Let $[T_{ij}]$ be the matrix from Table 1.1, and let the sets $\{F, G, P, Q\}$ and $\{\mathcal{F}, \mathcal{G}, \mathcal{P}, \mathcal{Q}\}$ be indexed by $1, \ldots, 4$, in obvious order. For $i, j, k = 1, \ldots, 4$ we have, in obvious notation, $i \, \mathcal{E}_k \, j$ iff $T_{kj} = T_{ki} T_{ij}$. A calculation shows that $i \, \mathcal{E}_k \, j$ iff $i = j$ or $i = k$, i.e., encompassing holds only under trivial conditions.

The images of one family into the others are given by

$$(6.1) \qquad\qquad \mathcal{F}_{\mathcal{G}} = \mathcal{F}_{\mathcal{P}} = \mathcal{F}_{\mathcal{Q}} = \emptyset,$$

$$(6.2) \qquad\qquad \mathcal{G}_{\mathcal{F}} = \mathcal{G}_{\mathcal{P}} = \mathcal{G}_{\mathcal{Q}} = \mathcal{G},$$

$$(6.3) \qquad\qquad \mathcal{P}_{\mathcal{F}} = \mathcal{P}_{\mathcal{G}} = \mathcal{P}_{\mathcal{Q}} = \mathcal{P},$$

$$(6.4) \qquad\qquad \mathcal{Q}_{\mathcal{F}} = \mathcal{Q}_{\mathcal{G}} = \mathcal{Q}_{\mathcal{P}} = \mathcal{Q},$$

and all relevant reflecting sets are empty, since $T_{ij} T_{ji} \neq 1$ for $i \neq j$. Finally, notice that each of the families $\mathcal{G}$, $\mathcal{P}$ and $\mathcal{Q}$ provides an incomplete parametric encompassing of any of the other families and of $\mathcal{F}$, relative to any distribution from the other families or from $\mathcal{F}$. This follows from the fact that the pseudo-true location parameters always coincide. Instances of complete parametric encompassing result from this fact by restricting attention to families having only a free location parameter. □

Example 6.2. Reconsider Example 1.5.2. Let D be an element of $\mathcal{G}$, i.e., $D = G_\beta$ for some $\beta \in \Omega_{\mathcal{G}}$. From Property 4.3, $\mathcal{F} \, \mathcal{E}_D \, \mathcal{G}$ iff D is in the reflecting set $\mathcal{G}_{\mathcal{F}}^R$. This set is determined by the solutions of $\beta_{\alpha_\beta} = \beta$ with respect to β. Using (1.5.7)–(1.5.8), we have

$$(6.5) \qquad\qquad \beta_{\alpha_\beta} = \left(\psi' \big(\kappa(\beta_1) \big), \beta_2 \right).$$

Since the function $\psi' \circ \kappa$ has no fixed points, $\mathcal{G}_{\mathcal{F}}^R$ is empty and $\mathcal{F} \, \not\mathcal{E}_D \, \mathcal{G}$. Notice from (6.5) that $\mathcal{F}$ provides an incomplete parametric encompassing of $\mathcal{G}$ relative to any $D \in \mathcal{G}$.

Since also $\mathcal{F}_{\mathcal{G}}^R$ is empty, it follows immediately that $\mathcal{G} \not\varepsilon_D \mathcal{F}$ when $D \in \mathcal{F}$. From (1.5.7)–(1.5.8) it follows that

$$(6.6) \qquad \alpha_{\beta_\alpha} = \left(\kappa(\psi'(\alpha_1)), \exp\left[\psi(\alpha_1) - \psi\big(\kappa(\psi'(\alpha_1))\big) \right] \alpha_2 \right).$$

Since $\kappa \circ \psi'$ has no fixed points and ψ is monotonically increasing, it is also apparent from (6.6) that $\mathcal{G}$ does not provide an incomplete parametric encompassing of $\mathcal{F}$ relative to any $D \in \mathcal{F}$. $\qquad \square$

Example 6.3. Reconsider Example 1.5.3. Let D be the conditionally normal distribution on $(R, \mathcal{B})$, given X, with conditional mean $d_1 = d_1(X)$ and conditional variance d_2. The pseudo-true value of β relative to D and P is given by

$$(6.7) \qquad \beta_D = \begin{pmatrix} [E_P(X_G X_G')]^{-1} E_P(X_G d_1) \\ d_2 + E_P(d_1^2) - E_P(d_1 X_G')\, [E_P(X_G X_G')]^{-1} E_P(X_G d_1) \end{pmatrix},$$

by analogy to (1.5.21). A similar expression holds for α_D. We assume that d_1 and the components of X_F and X_G are in the space $L_2(R^l, P)$ and that $E_P(X_F X_F')$ and $E_P(X_G X_G')$ are non-singular. These conditions are necessary and sufficient to guarantee the existence and uniqueness of α_D and β_D. The expression for β_{α_D} is obtained from (6.7) by replacing d_1 by $X_F' \alpha_{1D}$ and d_2 by α_{2D}, in obvious notation. This gives

$$(6.8) \quad \beta_{\alpha_D} = \begin{pmatrix} [E_P(X_G X_G')]^{-1} E_P(X_G X_F')\, [E_P(X_F X_F')]^{-1} E_P(X_F d_1) \\ d_2 + E_P(d_1^2) - E_P(d_1 X_F')\, [E_P(X_F X_F')]^{-1} E_P(X_F X_G') \\ [E_P(X_G X_G')]^{-1} E_P(X_G X_F') \\ [E_P(X_F X_F')]^{-1} E_P(X_F d_1) \end{pmatrix}.$$

Comparing (6.7) with (6.8), it follows that $\mathcal{F}\, \mathcal{E}_D^P\, \mathcal{G}$ iff

$$(6.9) \qquad E_P(X_G d_1) = E_P(X_G X_F')\, [E_P(X_F X_F')]^{-1} E_P(X_F d_1).$$

Thus $\mathcal{F}\, \mathcal{E}_D^P\, \mathcal{G}$ iff the projection of d_1 onto the orthogonal complement of X_F is orthogonal to X_G. In statistical terms, $\mathcal{F}\, \mathcal{E}_D^P\, \mathcal{G}$ iff X_G and d_1 are uncorrelated, given X_F.

Given a vector $(x_1, \ldots, x_T)$ of realizations of X, denote the $T \times 1$ vector $\big(d_1(x_1), \ldots, d_1(x_T)\big)'$ by δ_1. Assuming full column rank of Ξ_F and Ξ_G, the conditional counterparts to (6.7) and (6.8) are given by

$$(6.10) \qquad \beta_D^* = \begin{pmatrix} (\Xi_G' \Xi_G)^{-1} \Xi_G' \delta_1 \\ d_2 + \frac{1}{T} \delta_1' \left[I - \Xi_G (\Xi_G' \Xi_G)^{-1} \Xi_G' \right] \delta_1 \end{pmatrix},$$

49

and

(6.11) $\beta^{*}_{\alpha^{*}_{D}} =$

$$\left(\begin{array}{c} (\Xi'_G \Xi_G)^{-1} \Xi'_G \Xi_F (\Xi'_F \Xi_F)^{-1} \Xi'_F \delta_1 \\ d_2 + \frac{1}{T} \delta'_1 \left[I - \Xi_F (\Xi'_F \Xi_F)^{-1} \Xi'_F \Xi_G (\Xi'_G \Xi_G)^{-1} \Xi'_G \Xi_F (\Xi'_F \Xi_F)^{-1} \Xi'_F \right] \delta_1 \end{array} \right),$$

respectively. Clearly, $\beta^{*}_D = \beta^{*}_{\alpha^{*}_D}$ iff

(6.12) $$\Xi'_G \delta_1 = \Xi'_G \Xi_F (\Xi'_F \Xi_F)^{-1} \Xi'_F \delta_1.$$

Thus, $\mathcal{F} \, \mathcal{E}^{*}_D \, \mathcal{G}$ iff the projection of δ_1 onto the orthogonal complement of the space spanned by the columns of Ξ_F is orthogonal to the space spanned by the columns of Ξ_G.   □

Example 6.4. Reconsider Example 1.5.4. Let D be the conditionally normal distribution on $(R, \mathcal{B})$, given X, with conditional mean $d_1 = d_1(X) \in L_2(R^l, P)$ and conditional variance d_2. Then, from the results of Examples 1.5.3 and 1.5.4, β_D exists and none of α_D, β_{α_D} and α_{β_D} exists. Hence $\mathcal{F} \, \mathcal{L}^P_D \, \mathcal{G}$ and $\mathcal{G} \, \mathcal{E}^P_D \, \mathcal{F}$. Similarly, $\mathcal{F} \, \mathcal{L}^{*}_D \, \mathcal{G}$ and $\mathcal{G} \, \mathcal{E}^{*}_D \, \mathcal{F}$.

Alternatively, let D be the conditionally lognormal distribution on $(R, \mathcal{B})$, given X, with conditional log-mean $d_1 = d_1(X) \in L_2(R^l, P)$ and conditional log-variance d_2. Then, β_D, α_D and β_{α_D} exist and α_{β_D} does not exist. It follows that $\mathcal{G} \, \mathcal{L}^P_D \, \mathcal{F}$ and $\mathcal{G} \, \mathcal{L}^{*}_D \, \mathcal{F}$. To see whether $\mathcal{F}$ encompasses $\mathcal{G}$ or not, observe that

(6.13) $$\alpha_D = \left(\begin{array}{c} [E_P(X_F X'_F)]^{-1} E_P(X_F d_1) \\ d_2 + E_P(d_1^2) - E_P(d_1 X'_F) [E_P(X_F X'_F)]^{-1} E_P(X_F d_1) \end{array} \right)$$

and
(6.14)

$$\beta_D = \left(\begin{array}{c} [E_P(X_G X'_G)]^{-1} E_P(X_G m_1) \\ E_P m_2 + E_P(m_1^2) - E_P(m_1 X'_G) [E_P(X_G X'_G)]^{-1} E_P(X_G m_1) \end{array} \right),$$

where

(6.15) $$m_1 = \exp(d_1 + \tfrac{1}{2} d_2),$$

(6.16) $$m_2 = \exp(2d_1 + 2d_2) - m_1^2.$$

50

Furthermore, β_{α_D} is obtained from (6.14) by replacing m_1 by

$$(6.17) \qquad m_{1F} = \exp(X'_F \alpha_{1D} + \tfrac{1}{2}\alpha_{2D})$$

and m_2 by

$$(6.18) \qquad m_{2F} = \exp(2X'_F \alpha_{1D} + 2\alpha_{2D}) - \exp(2X'_F \alpha_{1D} + \alpha_{2D}).$$

Eventually, $\beta_D = \beta_{\alpha_D}$ iff

$$(6.19) \qquad E_P(X_G m_1) = E_P(X_G m_{1F})$$

and

$$(6.20) \qquad E_P m_2 + E_P(m_1^2) = E_P m_{2F} + E_P(m_{1F}^2).$$

Thus, $\mathcal{F} \, \mathcal{E}_D^P \, \mathcal{G}$ iff the projection of m_1 onto the orthogonal complement of m_{1F} is orthogonal to X_G and the expectations of the raw second moments of D and F_{α_D} are equal.

Given a vector $(x_1, \ldots, x_T)$ of realizations of X, denote the $T \times 1$ vector $\big(d_1(x_1), \ldots, d_1(x_T)\big)'$ by δ_1. Furthermore, define the $T \times 1$ vectors

$$(6.21) \qquad \mu_1 = \big[\exp\big(d_1(x_t) + \tfrac{1}{2}d_2\big)\big],$$

$$(6.22) \qquad \mu_2 = \big[\exp\big(2d_1(x_t) + 2d_2\big) - \exp\big(2d_1(x_t) + d_2\big)\big].$$

The conditional counterparts to (6.13) and (6.14) are given by

$$(6.23) \qquad \alpha_D^* = \begin{pmatrix} (\Xi'_F \Xi_F)^{-1} \Xi'_F \delta_1 \\ \delta_2 + \tfrac{1}{T}\delta'_1 \big[I - \Xi_F(\Xi'_F \Xi_F)^{-1}\Xi'_F\big]\delta_1 \end{pmatrix},$$

$$(6.24) \qquad \beta_D^* = \begin{pmatrix} (\Xi'_G \Xi_G)^{-1} \Xi'_G \mu_1 \\ \tfrac{1}{T}\iota'\mu_2 + \tfrac{1}{T}\mu'_1 \big[I - \Xi_G(\Xi'_G \Xi_G)^{-1}\Xi'_G\big]\mu_1 \end{pmatrix},$$

while $\beta_{\alpha_D^*}^*$ is obtained from (6.24) by replacing μ_1 by

$$(6.25) \qquad \mu_{1F} = \big[\exp\big(x'_F(x_t)\alpha_{1D}^* + \tfrac{1}{2}\alpha_{2D}^*\big)\big]$$

and μ_2 by

$$(6.26) \qquad \mu_{2F} = \big[\exp\big(2x'_F(x_t)\alpha_{1D}^* + 2\alpha_{2D}^*\big) - \exp\big(2x'_F(x_t)\alpha_{1D}^* + \alpha_{2D}^*\big)\big].$$

It follows that $\beta_D^* = \beta_{\alpha_D^*}^*$ iff

(6.27)
$$\Xi_G' \mu_1 = \Xi_G' \Xi_F (\Xi_F' \Xi_F)^{-1} \Xi_F' \mu_{1F}$$

and

(6.28)
$$\iota' \mu_2 + \mu_1' \mu_1 = \iota' \mu_{1F} + \mu_{1F}' \mu_{1F}.$$

Thus, $\mathcal{F} \, \mathcal{E}_D^* \, \mathcal{G}$ iff the projection of μ_{1F} onto the orthogonal complement of the space spanned by the columns of Ξ_F is orthogonal to the space spanned by the columns of Ξ_G and the average raw second moments of D and $F_{\alpha_D^*}$ are equal. □

7. Conclusion

We have analyzed the properties of the encompassing relation and its implications for empirical model building. Encompassing involves the comparison of a pseudo-true value relative to the DGP and a pseudo-true value relative to a pseudo-true distribution relative to the DGP. As such, encompassing formalizes the idea that a model should be able to account for the results obtained by another model. A failure of a model to encompass a rival model reveals shortcomings of the former model as an approximation to the DGP. Conversely, the finding that a model encompasses another model corroborates the former and at the same time allows a reduction of the encompassed model. However, from the non-transitivity of the encompassing relation we conclude that encompassed models cannot simply be discarded. Any decision based on the encompassing principle to reduce the number of empirical models should follow from (partial) knowledge of the encompassing relation. Hence, the question of statistical inference with respect to this relation arises. This question is taken up in the next chapter.

Chapter 3

Testing the encompassing hypothesis

1. Introduction

The statistical inference concerning the encompassing relation is the main object of this chapter. In general, theoretical considerations of the real world phenomena under study should narrow the scope to plausible models. If then we are left with a number of competing models which all have a sound theoretical basis, it is only natural to conduct statistical inference in order to assess the relative merits of these models. The significance of the encompassing relation for empirical model building justifies this interest.

For a pair of parametric models $\mathcal{F}$ and $\mathcal{G}$ from some given class $\mathcal{M}$, we are interested in testing the encompassing hypothesis $H_{\mathcal{E}} : \mathcal{F} \; \mathcal{E}_D^P \; \mathcal{G}$, given a body of data on (Y, X), generated from the distribution π_D. This distribution factors into D, the true conditional distribution of Y, given X, and P, the true marginal distribution of X. The models $\mathcal{F}$ and $\mathcal{G}$ are tentative approximations to D. Typically, D and P are unknown since otherwise we could, at least in principle, determine whether $H_{\mathcal{E}}$ holds or not. The lack of knowledge of D and P is partly met by the data generated by them, and the inference is to be based on these data only. From the Glivenko-Cantelli theorem, it follows that in principle it is possible to identify D and P from an infinite number of independent drawings, and thus to determine unambiguously whether $H_{\mathcal{E}}$ holds or not. Faced with a limited data set, however, the outcome of any sensible inference is inherently uncertain. Nevertheless, it is of interest to design inferential procedures which give asymptotically an unambiguous and correct answer to the question whether $H_{\mathcal{E}}$ holds or not. We are only able to do so under a restrictive set of regularity conditions on D, P, $\mathcal{F}$ and $\mathcal{G}$.

The tests we propose are applications of the three main principles in classical parametric inference: the Wald principle (Wald [1943]), the score principle (Rao [1947]) and the likelihood ratio principle (Neyman–Pearson [1928]). Since they are derived within the pseudo-likelihood (or quasi-likelihood) framework, the distribution theory underlying the tests is a simple extension of well established results from the pseudo-maximum likelihood theory (see, e.g., White [1982], Gouriéroux–Monfort–Trognon [1984], Vuong [1989] and Gouriéroux–Monfort [1995]). The tests are of an asymptotic nature since in general they have correct size only asymptotically. The main forms of the Wald and score tests have already been proposed by Gouriéroux–Monfort [1995]. They obtained the limit distributions of the test statistics under the encompassing hypothesis in the more general context of dynamic models.

Testing for encompassing is related to two other approaches in parametric inference: the robust approach to nested hypothesis testing and the non-robust approach to non-nested hypothesis testing, which may both be viewed as extensions of the classical non-robust approach to nested hypothesis testing. The robust nested approach considers nested hypotheses directly in terms of pseudo-true values and constructs tests which are robust under misspecification. By contrast, the non-robust non-nested approach considers non-nested hypotheses and constructs tests which are generally non-robust under misspecification. From this point of view, the encompassing approach can be seen as the extension of the robust approach to nested hypotheses testing to robust non-nested hypotheses testing or, equivalently, as the robustification of the standard non-nested approach. As a consequence, all the tests we obtain are at most extensions of tests which previously appeared in the literature. This view is somewhat at variance with the one put forward by Mizon–Richard [1986], who present the encompassing approach as the integration of the literature on (non-robust) nested and non-nested hypothesis testing. In line with this different view, their definition of encompassing differs from ours, although in their analysis of linear regression models the notion of encompassing is explicitly linked with the implicit null hypothesis of standard tests.

This chapter is organized as follows. In Section 2, we relate the encompassing hypothesis to other statistical hypotheses, in an attempt to provide a unified framework in which to interpret related approaches in parametric inference and to discuss the relations and differences among them. In Section 3, we restate the encompassing hypothesis in terms of almost sure limits of three important pairs of statistics which relate to the Wald, the score and the likelihood ratio principles. The limit distributions of these statistics are derived in Section 4 under general conditions, that is, both under the encompassing hypothesis and under deviations from it, and some asymptotic equivalences are noted. In Section 5, we study the asymptotic behaviour of quadratic forms in asymptotically normal variates. In Section 6, based on the previous results, we derive a large class of Wald, score and likelihood ratio encompassing tests, which comprises many parametric tests available in the literature. The special cases of nested models and hypotheses in implicit form are dealt with separately in Section 7. Section 8 concludes this chapter.

2. Encompassing and other statistical hypotheses

Consider the framework described in Section 2 of Chapter 2. In this framework, parametric inference is concerned with hypotheses testing relative to D (the true conditional distribution of $Y|X$) and P (the true marginal distribution of X) which can be expressed by means of parametric families. We distinguish between three approaches: the classical (non-robust) approach to nested hypothesis testing (see, e.g., Neyman–Pearson [1928], Wilks [1938], Wald [1943] and Rao [1947]), the robust approach to nested hypothesis testing (see, e.g., Kent [1982] and White [1982]) and the (non-robust) approach to non-nested hypothesis testing (see, e.g., Cox [1961, 1962], Atkinson [1970], Pesaran [1974], Pesaran and Deaton [1978], Fisher–McAleer [1981], McAleer [1981], Davidson–MacKinnon [1982], Gouriéroux–Monfort–Trognon [1983]). It is our view that the theory of encompassing provides a unified framework for interpreting each of these approaches and for discussing the relations and differences among them. It also allows an unambiguous statement of the hypothesis under test and of the implicit null hypothesis associated with such a test. Finally, it offers a clear view of the distribution theory underlying the usual test procedures and, related to this, it prompts the integration of the robust approach to nested hypothesis testing and the non-robust approach to non-nested hypothesis testing. The statistical implementation of this idea is taken up in later sections. In the following, we discuss how the hypotheses in the different approaches are related to the encompassing hypothesis, which is stated as

$$(2.1) \qquad\qquad H_{\mathcal{E}} : \mathcal{F} \, \mathcal{E}_D^P \, \mathcal{G}$$

or, equivalently, as

$$(2.2) \qquad\qquad H_{\mathcal{E}} : \beta_D = \beta_{\alpha_D},$$

for some parametric models $\mathcal{F} = \{F_\alpha | \alpha \in \Omega_{\mathcal{F}}\}$ and $\mathcal{G} = \{G_\beta | \beta \in \Omega_{\mathcal{G}}\}$.

The classical theory of parametric inference is concerned with hypothesis testing relative to D, which is known to belong to a *given* parametric family of distributions, say $\mathcal{G}$. The assumption

$$(2.3) \qquad\qquad H_M : D \in \mathcal{G}$$

is referred to as the 'maintained hypothesis', $\mathcal{G}$ being the 'maintained model'. In this setting, hypothesis testing is concerned with the question whether some given theoretical restrictions on D are true, in addition to H_M. The hypothesis to be tested, usually called the null (hypothesis), can always be stated in terms of a nested model $\mathcal{F} \subset \mathcal{G}$, viz.

$$(2.4) \qquad\qquad H_0 : D \in \mathcal{F}.$$

Then, it follows from Property 2.4.5 that H_0, H_M and $H_{\mathcal{E}}$ are related by

$$(2.5) \qquad\qquad H_0 \,|\, H_M \iff H_{\mathcal{E}},$$

showing that, *given* H_M, the classical hypothesis in (2.4) is in fact an encompassing hypothesis. The form $H_{\mathcal{E}} : \beta_D = \beta_{\alpha_D}$ also makes it clear that H_0 is a hypothesis on the pseudo-true value β_D (which, given H_M, is the true value), and therefore on D.

Robust hypothesis testing is concerned with the question whether some given theoretical restrictions on the pseudo-true value associated with a given model $\mathcal{G}$ are true or not, under the assumption that the pseudo-true value exists. No additional maintained hypothesis is assumed here, so that $\mathcal{G}$ may be misspecified. This justifies the term 'robust', by which is meant robust against specification errors. The null is formulated in terms of a nested model $\mathcal{F} \subset \mathcal{G}$, viz.

$$(2.6) \qquad\qquad H_0' : G_{\beta_D} = F_{\alpha_D}.$$

Using Property 2.4.5, it follows that H_0' and $H_{\mathcal{E}}$ are equivalent, i.e.,

$$(2.7) \qquad\qquad H_0' \iff H_{\mathcal{E}}.$$

Non-nested hypothesis testing is concerned with the problem of testing a model $\mathcal{F}$ against a non-nested alternative $\mathcal{G}$, (i.e., $\mathcal{F} \not\subset \mathcal{G}$ and usually $\mathcal{G} \not\subset \mathcal{F}$). This approach is closely related to the problem of model choice, i.e., choosing the 'best' model among $\mathcal{F}$ and $\mathcal{G}$ according to some criterion. Although rarely explicitly stated, it seems to be implicit in the literature on non-nested testing that one wants to test the null hypothesis

$$(2.8) \qquad\qquad H_0'' : D \in \mathcal{F}$$

under the maintained hypothesis

$$(2.9) \qquad\qquad H_M'' : D \in \mathcal{F} \cup \mathcal{G}.$$

The counterpart to (2.5) in the context of non-nested models is more complicated. First, recall the definition of the reflecting set $\mathcal{G}_{\mathcal{F}}^R = \{G_\beta \in \mathcal{G} \,|\, \exists\, F_\alpha \in \mathcal{F} : \beta = \beta_\alpha \text{ and } \alpha = \alpha_\beta\}$. Then, defining the stronger maintained hypothesis

$$(2.10) \qquad\qquad \bar{H}_M'' : D \in \mathcal{F} \cup (\mathcal{G} \setminus \mathcal{G}_{\mathcal{F}}^R),$$

it follows from Property 2.4.3 that

$$(2.11) \qquad\qquad H_0'' \,|\, \bar{H}_M'' \iff H_\varepsilon.$$

The usual procedure for testing each of the preceding types of hypotheses is to derive a probabilistic statement, based on the null and on the appropriate maintained hypothesis, concerning a sample of observations generated by D and P. Such a statement defines the critical region of the sample space and thereby a statistical test of the null. Since the null implies correct specification of $\mathcal{F}$ in the classical and the standard non-nested approaches, the distribution theory underlying these statements relies on the assumption that $\mathcal{F}$ is correctly specified. By contrast, the distribution theory invoked by the robust approach derives directly from D and P. Assuming certain regularity conditions, for each of the approaches it is possible to construct, for any given asymptotic level, consistent tests against any departure from the null, given the appropriate maintained hypothesis. As we will clarify later on, if no maintained hypothesis is assumed, a large class of such tests is consistent only against departures from H_ε. Hence, the implicit null associated with this class of tests is H_ε, which is usually much larger than the nominal null since in any case the null implies $H_{\mathcal{E}}$ irrespective of the maintained hypothesis—see Property 2.4.2.

Looking at the different approaches described above, it is a natural question whether the robust approach can be extended to non-nested hypothesis testing. Compared to the non-robust approach to non-nested hypothesis testing, this leads to the general question whether $H_{\mathcal{E}}$ holds or not, without assuming a maintained hypothesis like (2.3), (2.9) or (2.10) and without assuming that $\mathcal{F} \subset \mathcal{G}$. As a matter of fact, this formulation disconnects a hypothesis on the pseudo-true values from the maintained hypothesis and drops the latter. As we will show later on, it is possible to construct tests which are consistent against any departure from $H_{\mathcal{E}}$, provided that certain regularity conditions hold. The probabilistic statements underlying such tests are to be based solely on the encompassing hypothesis. Hence, the distribution theory has to be derived directly from the unknown distributions D and P, without assuming that $\mathcal{F}$ is correctly specified. The implicit null associated with such a test coincides with $H_{\mathcal{E}}$.

Figure 1 gives an overview of the different approaches in parametric inference discussed above. The lower-right part of the figure is the integration of robust and non-nested hypothesis testing: it extends the robust nested hypothesis testing approach to the non-nested case or, equivalently, it robustifies the standard non-nested hypothesis testing approach. Testing for encompassing may be seen as robust hypothesis testing, embracing nested as well as non-nested hypotheses.

Figure 1. *Nested, non-nested, robust and non-robust hypothesis testing.*

Nested, non-robust	**Non-nested, non-robust**
Models: $\mathcal{F} \subset \mathcal{G}$	Models: $\mathcal{F} \not\subset \mathcal{G}$
Maintained hypothesis: $D \in \mathcal{G}$	Maintained hypothesis: $D \in \mathcal{F} \cup (\mathcal{G} \setminus \mathcal{G}_{\mathcal{F}}^R)$
Null hypothesis: $D \in \mathcal{F}$	Null hypothesis: $D \in \mathcal{F}$
Distribution theory under: $\mathcal{F}, P$	Distribution theory under: $\mathcal{F}, P$
Nested, robust	**Non-nested, robust**
Models: $\mathcal{F} \subset \mathcal{G}$	Models: $\mathcal{F} \not\subset \mathcal{G}$
Maintained hypothesis: none	Maintained hypothesis: none
Null hypothesis: $\mathcal{F}\ \mathcal{E}_D^P\ \mathcal{G}$	Null hypothesis: $\mathcal{F}\ \mathcal{E}_D^P\ \mathcal{G}$
Distribution theory under: D, P	Distribution theory under: D, P

3. The encompassing hypothesis in terms of almost sure limits

The pseudo-ML (or quasi-ML) theory is the theory of ML estimation of possibly misspecified models. In this and the following sections, we give a brief account of some basic results of this theory, and show how a theory of robust hypothesis testing can be developed from these results. Here we give conditions under which the pseudo-ML estimator converges almost surely to its pseudo-true value. This property allows the encompassing hypothesis to be restated in terms of almost sure limits of some important statistics which are based on pseudo-ML estimators or some functions thereof.

The general framework may be described as follows. Let (C, σ_C, P) be a probability space, where C is the Euclidean space R^l and σ_C is the completion of the Borel σ-field on C with respect to the probability measure P. Let X be a random vector having distribution P, and taking values x in C. Let (A, σ_A, ν) be a σ-finite measure space, where A is the Euclidean space R^k, σ_A is the Borel σ-field on A and ν is a σ-finite measure on (A, σ_A). Let D and the elements of the parametric families $\mathcal{F} = \{F_\alpha | \alpha \in \Omega_{\mathcal{F}} \subset R^m\}$ and $\mathcal{G} = \{G_\beta | \beta \in \Omega_{\mathcal{G}} \subset R^n\}$ be conditional distributions on (A, σ_A), given X. Let Y be a random vector taking values y in A and whose conditional distribution, given X, is D. Finally, let π_D be the product probability measure defined by D and P on the product σ-field in $A \times C$ generated by $\sigma_A \times \sigma_C$.

The following subsection gives some basic almost sure convergence results. They concern the limit behaviour of the pseudo-ML estimators associated with $\mathcal{F}$ and $\mathcal{G}$ and some functions thereof. All these statistics are based on a sample of independent observations on Y and X which are generated by π_D. The main step in achieving the results is the application of the Kolmogorov strong law of large numbers. Subsequently, we define Wald and score vectors and a modified log-likelihood ratio and show how these relate to the encompassing hypothesis.

3.1. Basic convergence results

The measures ν, D and P and the families of measures $\mathcal{F}$ and $\mathcal{G}$ are assumed to satisfy the following regularity assumptions. Admittedly, the assumptions in this and the following sections are restrictive, in particular the differentiability and integrability conditions. The results probably hold under weaker assumptions. However, this would require different methods of proof and is likely to have a cost in terms of transparency. The search for sufficient regularity conditions which are closer to necessary is left to future work.

Throughout this chapter, any time we make an assumption with respect to $\mathcal{F}$, we implicitly make an analogous assumption with respect to $\mathcal{G}$.

Assumption A1. *For P-almost all x, D has a conditional density d relative to ν, given x.*

Assumption A2. *(a) For every $\alpha \in \Omega_{\mathcal{F}}$ and for P-almost all x, F_α has a conditional density f_α relative to ν, given x. (b) For π_D-almost all (y, x), f_α is continuous in α. (c) For every $\alpha \in \Omega_{\mathcal{F}}$ and for P-almost all x, ν is absolutely continuous with respect to F_α.*

Denote the conditional density of G_β relative to ν, given x, by g_β. Given Assumptions A1–A2, for P-almost all x, f_α and g_β have identical support for each $\alpha \in \Omega_{\mathcal{F}}$ and each $\beta \in \Omega_{\mathcal{G}}$. Moreover, this support does not depend on x and contains the support of d.

Assumption A3. *(a) $\Omega_{\mathcal{F}}$ is an open set. (b) α has a unique pseudo-true value α_D relative to D and P.*

Denote the unique pseudo-true value of β relative to D and P by β_D.

Assumption A4. *In some open neighbourhood of α_D, $|\log f(y|x; \alpha)|$ is dominated by a π_D-integrable function independent of α.*

Assume that we are given a sample of observations (y_t, x_t), $t = 1, \ldots, T$, which are generated independently by π_D. The normalized (pseudo-)log-likelihood functions associated with $\mathcal{F}$ and $\mathcal{G}$ for the sample are given by

$$(3.1) \qquad L_F(\alpha) = \frac{1}{T} \sum_{t=1}^{T} L_F^t(\alpha),$$

$$(3.2) \qquad L_G(\beta) = \frac{1}{T} \sum_{t=1}^{T} L_G^t(\beta),$$

respectively, where

$$(3.3) \qquad L_F^t(\alpha) = \log f(y_t | x_t; \alpha),$$

$$(3.4) \qquad\qquad L_G^t(\beta) = \log g(y_t|x_t; \beta).$$

The pseudo-ML estimators $\hat{\alpha}$ and $\hat{\beta}$, associated with $\mathcal{F}$ and $\mathcal{G}$, respectively, are defined as any solution of

$$(3.5) \qquad\qquad \max_{\alpha \in \Omega_{\mathcal{F}}} L_F(\alpha)$$

and

$$(3.6) \qquad\qquad \max_{\beta \in \Omega_{\mathcal{G}}} L_G(\beta),$$

respectively. For fixed T, $\hat{\alpha}$ and $\hat{\beta}$ need not exist, nor need they be unique. However, as $T \to \infty$, $\hat{\alpha}$ and $\hat{\beta}$ exist and are unique π_D-almost surely, and converge π_D-almost surely to α_D and β_D, respectively. To show this, let $\xrightarrow[\pi_D]{a.s.}$ denote π_D-almost sure convergence as $T \to \infty$. Then, applying the Kolmogorov strong law of large numbers yields the following:

$$(3.7) \qquad\qquad L_F(\alpha) \xrightarrow[\pi_D]{a.s.} E_P E_D \log f(Y|X; \alpha),$$

$$(3.8) \qquad\qquad L_G(\beta) \xrightarrow[\pi_D]{a.s.} E_P E_D \log g(Y|X; \beta),$$

whenever the expectations exist. It follows from Assumptions A3–A4 that the expectations exist in an open neighbourhood of their respective unique maximizers α_D and β_D. Hence, given the continuity of f in α and g in β,

$$(3.9) \qquad\qquad \hat{\alpha} \xrightarrow[\pi_D]{a.s.} \alpha_D,$$

$$(3.10) \qquad\qquad \hat{\beta} \xrightarrow[\pi_D]{a.s.} \beta_D,$$

—see also White [1981, 1982].

We impose the following standard smoothness and integrability conditions. The matrix (or vector) whose elements are the absolute values of the corresponding elements of the matrix (or vector) M is denoted by $|M|$.

Assumption A5. *For π_D-almost all (y, x), $\log f(y|x; \alpha)$ is twice continuously differentiable on $\Omega_{\mathcal{F}}$.*

Assumption A6. *In some open neighbourhood of α_D, the components of $|\partial \log f(y|x; \alpha)/\partial \alpha|$ are dominated by π_D-integrable functions independent of α.*

Assumption A5 ensures the existence of the normalized score functions

$$(3.11) \qquad Q_F(\alpha) = \frac{1}{T} \sum_{t=1}^{T} Q_F^t(\alpha),$$

$$(3.12) \qquad Q_G(\beta) = \frac{1}{T} \sum_{t=1}^{T} Q_G^t(\beta),$$

where

$$(3.13) \qquad Q_F^t(\alpha) = \frac{\partial L_F^t(\alpha)}{\partial \alpha},$$

$$(3.14) \qquad Q_G^t(\beta) = \frac{\partial L_G^t(\beta)}{\partial \beta},$$

and of their derivatives

$$(3.15) \qquad H_F(\alpha) = \frac{1}{T} \sum_{t=1}^{T} H_F^t(\alpha),$$

$$(3.16) \qquad H_G(\beta) = \frac{1}{T} \sum_{t=1}^{T} H_G^t(\beta),$$

where

$$(3.17) \qquad H_F^t(\alpha) = \frac{\partial Q_F^t(\alpha)}{\partial \alpha'},$$

$$(3.18) \qquad H_G^t(\beta) = \frac{\partial Q_G^t(\beta)}{\partial \beta'}.$$

Furthermore, upon existence, $\hat{\alpha}$ and $\hat{\beta}$ satisfy the first order conditions

(3.19) $$Q_F(\hat{\alpha}) = 0$$

and

(3.20) $$Q_G(\hat{\beta}) = 0,$$

respectively. If in addition Assumption A6 holds, α_D and β_D satisfy the first order conditions

(3.21) $$E_P E_D \, Q_F^t(\alpha_D) = 0$$

and

(3.22) $$E_P E_D \, Q_G^t(\beta_D) = 0,$$

respectively.

We will sometimes need the following assumption. It ensures that the solution of the first order condition corresponding to the maximum of $E_P E_D \log g(Y|X;\beta)$ is unique.

Assumption A7. $E_P E_D \left[\partial \log g(Y|X;\beta)/\partial \beta \right] = 0$ only if $\beta = \beta_D$.

For a given $\alpha \in \Omega_{\mathcal{F}}$, denote the pseudo-true value of β relative to F_α and P by β_α. Recall that β_α maximizes $E_P E_\alpha \log g(Y|X;\beta)$ with respect to β. Since encompassing involves the comparison of β_D and β_{α_D}, we need to extend the previous discussion of the pseudo-true values and their estimators to the study of the pseudo-true value *mapping* β_α and the estimation of the values of this mapping. We impose regularity conditions on the behaviour of β_α in the neighbourhood of α_D. In the sequel, π_α denotes the product probability measure defined by F_α and P on the product σ-field in $A \times C$ generated by $\sigma_A \times \sigma_C$. The following two assumptions are similar to Assumptions A3(b)–A4.

Assumption B1. For all α in some open neighbourhood of α_D, β has a unique pseudo-true value β_α relative to F_α and P.

Assumption B2. For all α in some open neighbourhood of α_D, there exists an open neighbourhood of β_{α_D} where $|\log g(y|x;\beta)|$ is dominated by a π_α-integrable function independent of β.

By definition, $\beta_{\hat{\alpha}}$ is any solution of

(3.23) $$\max_{\beta \in \Omega_G} \left[E_P E_\alpha \log g(Y|X;\beta) \right]_{\alpha = \hat{\alpha}}.$$

63

As noted by Gouriéroux–Monfort–Trognon [1983] and Gouriéroux–Monfort [1995], unless $\mathcal{F} \subset \mathcal{G}$, the function β_α depends on P, which is generally unknown. It is therefore of interest also to consider the conditional pseudo-true value of β relative to F_α, given $(x_1, \ldots, x_T)$, denoted by β_α^*. Recall that β_α^* maximizes $T^{-1} \sum_{t=1}^{T} E_\alpha \log g(Y|x_t; \beta)$ with respect to β. Hence, β_α^* can always be computed numerically, if not analytically, from the sample (y_t, x_t), $t = 1, \ldots, T$. $\beta_{\hat{\alpha}}^*$ is then any solution of

$$(3.24) \qquad \max_{\beta \in \Omega_{\mathcal{G}}} \frac{1}{T} \sum_{t=1}^{T} [E_\alpha \log g(Y|x_t; \beta)]_{\alpha = \hat{\alpha}} .$$

Note as before that, for fixed T, $\beta_{\hat{\alpha}}$ and $\beta_{\hat{\alpha}}^*$ need not exist, nor need they be unique. An extension of the above argument shows that, as $T \to \infty$, $\beta_{\hat{\alpha}}$ and $\beta_{\hat{\alpha}}^*$ exist and are unique π_D-almost surely, and converge π_D-almost surely to β_{α_D}. From the Kolmogorov strong law of large numbers, it follows that

$$(3.25) \qquad \frac{1}{T} \sum_{t=1}^{T} E_{\alpha_D} \log g(Y|x_t; \beta) \xrightarrow[P]{a.s.} E_P E_{\alpha_D} \log g(Y|X; \beta),$$

and hence, given Assumption B2, that

$$(3.26) \qquad [E_P E_\alpha \log g(Y|X; \beta)]_{\alpha = \hat{\alpha}} \xrightarrow[\pi_D]{a.s.} E_P E_{\alpha_D} \log g(Y|X; \beta),$$

$$(3.27) \qquad \frac{1}{T} \sum_{t=1}^{T} [E_\alpha \log g(Y|x_t; \beta)]_{\alpha = \hat{\alpha}} \xrightarrow[\pi_D]{a.s.} E_P E_{\alpha_D} \log g(Y|X; \beta),$$

whenever the expectations exist. It follows from Assumptions B1–B2 that the expectations exist in an open neighbourhood of the unique maximizer β_{α_D}. Hence, from the continuity of g in β,

$$(3.28) \qquad \beta_{\hat{\alpha}} \xrightarrow[\pi_D]{a.s.} \beta_{\alpha_D},$$

$$(3.29) \qquad \beta_{\hat{\alpha}}^* \xrightarrow[\pi_D]{a.s.} \beta_{\alpha_D} .$$

We note that (3.24) may be difficult to solve in practice. The maximand involves an expectation for each t which frequently does not have a closed form. In this case, each evaluation of the maximand requires T

numerical (one-dimensional) integrations. Hence, if β is large-dimensional, solving (3.24) may be quite demanding. To avoid this computational burden, Gouriéroux–Monfort [1995] propose to estimate $\beta_{\hat{\alpha}}^*$ by means of Monte Carlo methods, that is, by the solution of

$$(3.30) \qquad \max_{\beta \in \Omega_G} \frac{1}{T} \sum_{t=1}^{T} \log g(y_t^s(\hat{\alpha})|x_t; \beta),$$

where $y_t^s(\hat{\alpha})$, $t = 1, \ldots, T$, are independent drawings from $F_{\hat{\alpha}}$, given x_t. Denoting the solution of (3.30) by $\beta_{\hat{\alpha}}^s$, it is not difficult to see that $E(\beta_{\hat{\alpha}}^s|\hat{\alpha}) = \beta_{\hat{\alpha}}^*$ (where the expectation is taken with respect to the Monte Carlo simulation), and that $\beta_{\hat{\alpha}}^s \xrightarrow[\pi_D]{a.s.} \beta_{\alpha_D}$. Obviously, taking the average of a number of such simulated pseudo-true values instead of only one would yield a more precise estimator of $\beta_{\hat{\alpha}}^*$, but a single simulation is sufficient to ensure consistency. Though having the same almost sure limit as $\beta_{\hat{\alpha}}^*$, the covariance matrix of the limit distribution of $\beta_{\hat{\alpha}}^s$ generally exceeds that of $\beta_{\hat{\alpha}}^*$ due to the extra randomness caused by the simulation. This complication is beyond the scope of the present study, where we shall only be concerned with tests based on $\beta_{\hat{\alpha}}^*$. See Gouriéroux–Monfort [1995] for encompassing tests based on simulated pseudo-true values.

The following smoothness and integrability assumptions related to $\log g$ are similar to Assumptions A5–A6.

Assumption B3. *For π_{α_D}-almost all (y, x), $\log g(y|x; \beta)$ is twice continuously differentiable on Ω_G.*

Assumption B4. *For all α in some open neighbourhood of α_D, there exists an open neighbourhood of β_{α_D} where the components of $|\partial \log g(y|x; \beta)/\partial \beta|$ are dominated by π_α-integrable functions independent of β.*

Given Assumptions A1–A6 and B1–B3, whenever $\beta_{\hat{\alpha}}$ and $\beta_{\hat{\alpha}}^*$ exist, they satisfy the first order conditions

$$(3.31) \qquad \left[E_P E_\alpha \left[Q_G^t(\beta) \right] \right]_{\alpha=\hat{\alpha}, \beta=\beta_{\hat{\alpha}}} = 0$$

and

$$(3.32) \qquad \frac{1}{T} \sum_{t=1}^{T} \left[E_\alpha \left[Q_G^t(\beta) \right] \right]_{\alpha=\hat{\alpha}, \beta=\beta_{\hat{\alpha}}^*} = 0,$$

respectively. If, moreover, Assumption B4 holds, β_α satisfies the first order condition

$$(3.33) \qquad\qquad E_P E_\alpha Q_G^t(\beta_\alpha) = 0$$

in some open neighbourhood of α_D.

The following assumptions guarantee the existence of $E_P E_D Q_G(\beta_{\alpha_D})$ and $E_P E_D L_G(\beta_{\alpha_D})$. Furthermore, they ensure that these expectations are the π_D-almost sure limits of $Q_G(\beta_{\hat\alpha})$ and $Q_G(\beta_{\hat\alpha}^*)$, and of $L_G(\beta_{\hat\alpha})$ and $L_G(\beta_{\hat\alpha}^*)$, respectively. As they are stated, the assumptions are stronger than necessary for the purposes mentioned, but they will be needed to justify the Taylor expansions used later on.

Assumption B5. Ω_G *is a convex set.*

Assumption B6. *In some open set containing the line segment joining β_D and β_{α_D}, the components of $|\partial \log g(y|x;\beta)/\partial\beta|$ are dominated by π_D-integrable functions independent of β.*

Assumption B7. *In some open set containing the line segment joining β_D and β_{α_D}, $|\log g(y|x;\beta)|$ is dominated by a π_D-integrable function independent of β.*

3.2. Definition of the basic statistics

We now want to reformulate the encompassing hypothesis in terms of π_D-almost sure limits of some important statistics of the data. Note that the pseudo-likelihood theory with respect to $\mathcal{G}$ is concerned with the optimization of the π_D-almost surely convergent random function $L_G(\beta)$ with respect to β. In such a setting, the classical theory on parametric inference usually proceeds along one of three main lines. Each of these in essence compares the unconstrained optimization of L_G with the constrained optimization of L_G under the hypothesis to be tested. The Wald principle compares the unconstrained and the constrained optimizers, the score principle compares the unconstrained and the constrained gradients (or scores) of the objective function, and the likelihood principle compares the unconstrained and the constrained optima. In each instance, the difference between the unconstrained and the constrained quantities converges to zero if and only if the hypothesis to be tested is true. A similar approach can be adopted for testing the encompassing hypothesis. For this purpose, we define the differences

$$(3.34) \qquad \phi = \beta_D - \beta_{\alpha_D},$$

$$(3.35) \qquad \hat{\phi} = \hat{\beta} - \beta_{\hat{\alpha}},$$

$$(3.36) \qquad \hat{\phi}^* = \hat{\beta} - \beta_{\hat{\alpha}}^*,$$

the scores

$$(3.37) \qquad q = E_P E_D \, Q_G(\beta_{\alpha_D}),$$

$$(3.38) \qquad \hat{q} = Q_G(\beta_{\hat{\alpha}}),$$

$$(3.39) \qquad \hat{q}^* = Q_G(\beta_{\hat{\alpha}}^*),$$

and the modified log-likelihood ratios

$$(3.40) \qquad l = E_P E_D \left[L_G(\beta_{\alpha_D}) - L_G(\beta_D) \right],$$

$$(3.41) \qquad \hat{l} = L_G(\beta_{\hat{\alpha}}) - L_G(\hat{\beta}),$$

$$(3.42) \qquad \hat{l}^* = L_G(\beta_{\hat{\alpha}}^*) - L_G(\hat{\beta}).$$

Obviously, $\hat{\phi}$ and $\hat{\phi}^*$ converge π_D-almost surely to ϕ, $\hat{q}$ and $\hat{q}^*$ converge π_D-almost surely to q, and $\hat{l}$ and $\hat{l}^*$ converge π_D-almost surely to l. Some further remarks are in order. First, whenever the determination of $\beta_{\hat{\alpha}}$ requires knowledge of P, the statistics $\hat{\phi}$, $\hat{q}$ and $\hat{l}$ cannot be computed from the sample (y_t, x_t), $t = 1, \ldots, T$, alone, whereas the statistics $\hat{\phi}^*$, $\hat{q}^*$ and $\hat{l}^*$ always can. Hence, inferential procedures regarding the encompassing hypothesis based on $\hat{\phi}$, $\hat{q}$ and $\hat{l}$ are often unfeasible—see also Gouriéroux–Monfort [1995]. As a consequence, the study of $\hat{\phi}^*$, $\hat{q}^*$ and $\hat{l}^*$ seems to be imperative. We shall characterize the asymptotic behaviour of $\hat{\phi}$, $\hat{q}$ and $\hat{l}$, and of $\hat{\phi}^*$, $\hat{q}^*$ and $\hat{l}^*$ under general conditions in Section 4. As will be shown, the results for the latter set of statistics are more complicated and somewhat harder to obtain. Secondly, the (normalized) log-likelihood ratio statistic is usually defined as $LR = L_F(\hat{\alpha}) - L_G(\hat{\beta})$. This statistic is identical to $\hat{l}$ and to $\hat{l}^*$ whenever $\mathcal{F} \subset \mathcal{G}$, but generally differs from $\hat{l}$ and from $\hat{l}^*$ when $\mathcal{F} \not\subset \mathcal{G}$. As far as I know, when $\mathcal{F} \not\subset \mathcal{G}$ the literature has only been concerned with the study of LR. In this respect, it is worth mentioning the asymptotic normality results of Cox [1961] and Vuong [1989] in connection with LR. Cox [1961] showed that LR minus its expected value under $\mathcal{F}$ is asymptotically normally distributed if $\mathcal{F}$ is correctly specified. Vuong [1989], in a model selection approach, proved that LR has a limiting weighted sum of chi-squares distribution if the pseudo-true distributions of $\mathcal{F}$ and $\mathcal{G}$ relative to D and P are identical, and a limiting normal distribution otherwise. Here, we follow a suggestion made by Gouriéroux–Monfort–Trognon [1983] and study the *modified* log-likelihood ratio statistics $\hat{l}$ and $\hat{l}^*$. It is worth noting that l is equal to the degree of non-encompassing defined in Chapter 2, i.e., $l = \mathcal{NE}_D^P(\mathcal{F}, \mathcal{G})$. It has been argued by, for example, Mizon–Richard [1986] that the asymptotic equivalence of some Wald and score encompassing tests (based on $\hat{\phi}$ or $\hat{\phi}^*$ and on $\hat{q}$ or $\hat{q}^*$, respectively) and the likelihood ratio test (based on LR), which holds when $\mathcal{F} \subset \mathcal{G}$, cannot be obtained when $\mathcal{F} \not\subset \mathcal{G}$. We will show that by replacing LR by $\hat{l}$ or $\hat{l}^*$ it is possible to obtain asymptotically equivalent Wald, score and likelihood ratio encompassing tests. Hence, whereas LR is appropriate in a model selection approach, the modified likelihood ratio statistics suit better in the encompassing framework. As a matter of fact, the modified likelihood ratio statistic has been proposed and studied independently by Dhaene [1993], which constitutes the predecessor of this monograph, and Smith [1993].

We have the following lemma.

Lemma 3.1.

(i) Given Assumptions A1–A4 and B1–B2,

$$(3.43) \qquad \mathcal{F} \, \mathcal{E}_D^P \, \mathcal{G} \iff \phi = 0 \iff \hat{\phi} \xrightarrow[\pi_D]{a.s.} 0 \iff \hat{\phi}^* \xrightarrow[\pi_D]{a.s.} 0.$$

(ii) Given Assumptions A1–A7 and B1–B6,

$$(3.44) \qquad \mathcal{F} \, \mathcal{E}_D^P \, \mathcal{G} \iff q = 0 \iff \hat{q} \xrightarrow[\pi_D]{a.s.} 0 \iff \hat{q}^* \xrightarrow[\pi_D]{a.s.} 0.$$

(iii) Given Assumptions A1–A4, B1–B2, B5 and B7,

$$(3.45) \qquad \mathcal{F} \, \mathcal{E}_D^P \, \mathcal{G} \iff l = 0 \iff \hat{l} \xrightarrow[\pi_D]{a.s.} 0 \iff \hat{l}^* \xrightarrow[\pi_D]{a.s.} 0.$$

Proof. The equivalences follow immediately from the π_D-almost sure convergence of $\hat{\beta}$, $\hat{\beta}_{\hat{\alpha}}$ and $\beta_{\hat{\alpha}}^*$ to β_D, β_{α_D} and β_{α_D}, respectively, and from the π_D-almost sure convergence of the score and the likelihood functions in the neighbourhoods of the pseudo-true values. $\qquad\qquad\square$

Encompassing tests can thus be based on the sample statistics $\hat{\phi}$ or $\hat{\phi}^*$, $\hat{q}$ or $\hat{q}^*$, and $\hat{l}$ or $\hat{l}^*$, giving rise to Wald, score and likelihood ratio encompassing tests, respectively.

Figure 1 provides a geometric interpretation of the quantities ϕ, q and l when β is one-dimensional. It depicts the functions $E_P E_D L_G^t(\beta)$ and $E_P E_{\alpha_D} L_G^t(\beta)$, which, by definition, are maximized by β_D and β_{α_D}, respectively. If $\mathcal{F} \not\subset_D^P \mathcal{G}$, as in the figure, then these maximizers do not coincide. The quantities ϕ, q and l are defined by the points β_D and β_{α_D} and by the properties of the function $E_P E_D L_G^t(\beta)$ at these points. They are given by the difference between β_D and β_{α_D}, the difference between the slopes of $E_P E_D L_G^t(\beta)$ at β_{α_D} and β_D, and the difference between the values of $E_P E_D L_G^t(\beta)$ at β_{α_D} and β_D, respectively. Note that $\text{sign}\,\phi = \text{sign}\,q$ and $l < 0$ whenever $\mathcal{F} \not\subset_D^P \mathcal{G}$. If $\mathcal{F} \subset_D^P \mathcal{G}$, then the maximizers of $E_P E_D L_G^t(\beta)$ and $E_P E_{\alpha_D} L_G^t(\beta)$ coincide. Finally, if $D \in \mathcal{F}$, then these functions themselves coincide.

Figure 1. *Geometric interpretation of ϕ, q and l.*

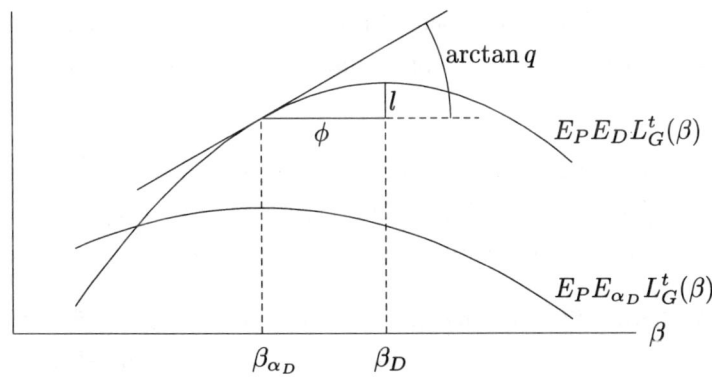

4. Limit distributions of the basic statistics

As a prerequisite to the construction of encompassing tests, we derive the limit distributions of $\hat{\phi}$, $\hat{\phi}^*$, $\hat{q}$, $\hat{q}^*$, $\hat{l}$ and $\hat{l}^*$ under general conditions. By this we mean that the models $\mathcal{F}$ and $\mathcal{G}$ may be nested, disjoint or overlapping, that both, only one or neither may be correctly specified and that both, only one or neither may encompass the other. We show that, under further regularity conditions, $\hat{\phi}$, $\hat{\phi}^*$, $\hat{q}$, and $\hat{q}^*$ are asymptotically normally distributed. On the other hand, the limit distributions of $\hat{l}$ and $\hat{l}^*$ are shown to be weighted sum of chi-squares distributions or normal distributions, depending on whether $\mathcal{F} \, \mathcal{E}_D^P \, \mathcal{G}$ or $\mathcal{F} \, \mathcal{E}_D^P \, \mathcal{G}$. The proofs proceed along standard lines. Taylor expansions are used to obtain approximations of the score and likelihood functions by means of sums of independent identically distributed random vectors on which the multivariate central limit theorem can be applied.

The following subsection presents some preliminary results. It gives some relations between ϕ, q and l, and establishes the asymptotic normality of $\beta_{\hat{\alpha}}$ and $\beta_{\hat{\alpha}}^*$. The second subsection derives the limit distributions of $\hat{\phi}$, $\hat{\phi}^*$, $\hat{q}$, $\hat{q}^*$, $\hat{l}$ and $\hat{l}^*$ from these results. Some asymptotic equivalences between $\hat{\phi}$, $\hat{q}$ and $\hat{l}$, and between $\hat{\phi}^*$, $\hat{q}^*$ and $\hat{l}^*$ are noted in the third subsection.

4.1. Asymptotic normality: preliminary results

The following three assumptions ensure the joint asymptotic normality of $\hat{\alpha}$ and $\hat{\beta}$. As before, any assumption made with respect to $\mathcal{F}$ is also implicitly made with respect to $\mathcal{G}$.

Assumption N1. *In some open neighbourhood of α_D, the components of $|\partial^2 \log f(y|x; \alpha)/\partial\alpha\partial\alpha'|$ and $|\partial \log f(y|x; \alpha)/\partial\alpha \cdot \partial \log f(y|x; \alpha)/\partial\alpha'|$ are dominated by π_D-integrable functions independent of α.*

Assumption N2. *In some open set containing the line segment joining β_D and β_{α_D}, the components of $|\partial^2 \log g(y|x; \beta)/\partial\beta\partial\beta'|$ and $|\partial \log g(y|x; \beta)/\partial\beta \cdot \partial \log g(y|x; \beta)/\partial\beta'|$ are dominated by π_D-integrable functions independent of β.*

Before proceeding, it is of interest to note the following relations between ϕ, q and l.

Lemma 4.1. *Given Assumptions A1–A6, B1–B7 and N2,*

$$(4.1) \qquad l = \frac{1}{2}\phi' E_P E_D \left[H_G^t(\beta_*) \right] \phi,$$

$$(4.2) \qquad q = -E_P E_D \left[H_G^t(\beta_{**}) \right] \phi,$$

for some β_* and β_{**} on the line segment joining β_D and β_{α_D}.

Proof. From the mean value theorem,

$$(4.3) \quad E_P E_D \left[L_G^t(\beta_{\alpha_D}) \right] = E_P E_D \left[L_G^t(\beta_D) \right] + E_P E_D \left[Q_G^t(\beta_D) \right] (\beta_{\alpha_D} - \beta_D)$$
$$+ \frac{1}{2}(\beta_{\alpha_D} - \beta_D)' E_P E_D \left[H_G^t(\beta_*) \right] (\beta_{\alpha_D} - \beta_D),$$

$$(4.4) \qquad E_P E_D \left[Q_G^t(\beta_{\alpha_D}) \right] = E_P E_D \left[Q_G^t(\beta_D) \right]$$
$$+ E_P E_D \left[H_G^t(\beta_{**}) \right] (\beta_{\alpha_D} - \beta_D),$$

for some β_* and β_{**} on the line segment joining β_D and β_{α_D}. Using (3.22), the results follow. $\qquad\square$

As the above lemma shows, ϕ and q are linearly related, while l is a quadratic form in ϕ, and hence in q. We may therefore expect that, under general conditions, the asymptotic behaviour of $\hat{l}$ (resp. $\hat{l}^*$) will be similar to that of quadratic forms in $\hat{\phi}$ (resp. $\hat{\phi}^*$) or in $\hat{q}$ (resp. $\hat{q}^*$). The results of the following sections confirm this conjecture.

Assumptions N1–N2 ensure the existence of the following matrices:

$$(4.5) \qquad H_F = -E_P E_D \left[H_F^t(\alpha_D) \right],$$

$$(4.6) \qquad H_G = -E_P E_D \left[H_G^t(\beta_D) \right],$$

$$(4.7) \qquad J_F = E_P E_D \left[Q_F^t(\alpha_D) Q_F^{t\prime}(\alpha_D) \right],$$

$$(4.8) \qquad J_G = E_P E_D \left[Q_G^t(\beta_D) Q_G^{t\prime}(\beta_D) \right],$$

$$(4.9) \qquad J_{FG} = E_P E_D \left[Q_F^t(\alpha_D) Q_G^{t\prime}(\beta_D) \right] = J_{GF}',$$

$$(4.10) \qquad \tilde{H}_G = -E_P E_D \left[H_G^t(\beta_{\alpha_D}) \right],$$

$$(4.11) \qquad \tilde{J}_G = E_P E_D \left[Q_G^t(\beta_{\alpha_D}) Q_G^{t\prime}(\beta_{\alpha_D}) \right] - qq',$$

$$(4.12) \qquad \tilde{J}_{FG} = E_P E_D \left[Q_F^t(\alpha_D) Q_G^{t\prime}(\beta_{\alpha_D}) \right] = \tilde{J}_{GF}'.$$

Assumption N3. $E_P E_D \left[H_F^t(\alpha) \right]$ *and* $E_P E_D \left[Q_F^t(\alpha) Q_F^{t\prime}(\alpha) \right]$ *are non-singular in some open neighbourhood of α_D.*

Let $\xrightarrow[\pi_D]{d}$ denote convergence in distribution under π_D as $T \to \infty$.

Lemma 4.2. *Given Assumptions A1–A6, B5 and N1–N3,*

$$(4.13) \qquad \sqrt{T} \begin{pmatrix} \hat{\alpha} - \alpha_D \\ \hat{\beta} - \beta_D \end{pmatrix} \xrightarrow[\pi_D]{d} N \left(0, \begin{pmatrix} H_F^{-1} J_F H_F^{-1} & H_F^{-1} J_{FG} H_G^{-1} \\ H_G^{-1} J_{GF} H_F^{-1} & H_G^{-1} J_G H_G^{-1} \end{pmatrix} \right).$$

Proof. See Vuong [1989], Lemma A. □

The following two assumptions ensure the asymptotic normality of $\beta_{\hat{\alpha}}$ and $\beta_{\hat{\alpha}}^*$.

Assumption N4. *For all α in some open neighbourhood of α_D, there exists an open neighbourhood of β_{α_D} where the components of $|\partial^2 \log g(y|x; \beta)/ \partial\beta\partial\beta'|$ and $|\partial \log g(y|x; \beta)/\partial\beta \cdot \partial \log g(y|x; \beta)/\partial\beta'|$ are dominated by π_α-integrable functions independent of β.*

Assumption N4 ensures the existence of the following matrices:

$$(4.14) \qquad \bar{H}_G = -E_P E_{\alpha_D} \left[H_G^t (\beta_{\alpha_D}) \right],$$

$$(4.15) \qquad \bar{J}_{FG} = E_P E_{\alpha_D} \left[Q_F^t(\alpha_D) Q_G^{t\prime} (\beta_{\alpha_D}) \right] = \bar{J}_{GF}',$$

$$(4.16) \qquad J_{\tilde{G}} = E_P \left[E_{\alpha_D} \left[Q_G^t (\beta_{\alpha_D}) \right] E_{\alpha_D} \left[Q_G^{t\prime} (\beta_{\alpha_D}) \right] \right],$$

$$(4.17) \qquad J_{F\tilde{G}} = E_P \left[E_D \left[Q_F^t (\alpha_D) \right] E_{\alpha_D} \left[Q_G^{t\prime} (\beta_{\alpha_D}) \right] \right] = J_{\tilde{G}F}',$$

$$(4.18) \qquad J_{G\tilde{G}} = E_P \left[E_D \left[Q_G^t (\beta_D) \right] E_{\alpha_D} \left[Q_G^{t\prime} (\beta_{\alpha_D}) \right] \right] = J_{\tilde{G}G}',$$

$$(4.19) \qquad \tilde{J}_{G\tilde{G}} = E_P \left[E_D \left[Q_G^t (\beta_{\alpha_D}) \right] E_{\alpha_D} \left[Q_G^{t\prime} (\beta_{\alpha_D}) \right] \right] = \tilde{J}_{\tilde{G}G}'.$$

From a computational point of view, it is important to note that the matrices defined in (4.5)–(4.12) and (4.14)–(4.19) are the π_D-almost sure limits of their sample analogues, obtained by replacing $E_P E_D$ by $T^{-1} \sum_{t=1}^T$, E_P by $T^{-1} \sum_{t=1}^T$, q by $\hat{q}^*$, α_D by $\hat{\alpha}$, β_D by $\hat{\beta}$, and β_{α_D} by $\beta_{\hat{\alpha}}^*$, successively.

Assumption N5. $E_P E_\alpha [H_G^t(\beta_\alpha)]$ is non-singular in some open neighbourhood of α_D.

Define

(4.20)
$$B(\alpha) = \frac{\partial \beta_\alpha}{\partial \alpha'}$$

and

(4.21)
$$B = B(\alpha_D).$$

Differentiating (3.33) with respect to α, evaluating the result at $\alpha = \alpha_D$ and solving for B gives

(4.22)
$$B = \bar{H}_G^{-1} \bar{J}_{GF}.$$

The following lemma establishes the asymptotic normality of $\beta_{\hat{\alpha}}$ and $\beta_{\hat{\alpha}}^*$.

Lemma 4.3. *Given Assumptions A1–A6, B1–B7 and N1–N5,*

(4.23)
$$\sqrt{T}(\beta_{\hat{\alpha}} - \beta_{\alpha_D}) \xrightarrow[\pi_D]{d} N(0, V_b),$$

(4.24)
$$\sqrt{T}(\beta_{\hat{\alpha}}^* - \beta_{\alpha_D}) \xrightarrow[\pi_D]{d} N(0, V_b^*),$$

where

(4.25)
$$V_b = B H_F^{-1} J_F H_F^{-1} B',$$

(4.26)
$$V_b^* = B H_F^{-1} J_F H_F^{-1} B' + \bar{H}_G^{-1} J_{\bar{G}} \bar{H}_G^{-1}$$
$$+ B H_F^{-1} J_{F\bar{G}} \bar{H}_G^{-1} + \bar{H}_G^{-1} J_{\bar{G}F} H_F^{-1} B'.$$

Proof. Expanding the likelihood equation $\sqrt{T} Q_F(\hat{\alpha}) = 0$ yields

(4.27)
$$0 = \sqrt{T} Q_F(\alpha_D) + \sqrt{T} H_F(\alpha_D)(\hat{\alpha} - \alpha_D) + o_p(1),$$

where $o_p(1)$ denotes a quantity converging to zero in probability under π_D. Hence, noting that $H_F(\alpha_D) \xrightarrow[\pi_D]{a.s.} H_F$ and that, from Lemma 4.2, $\sqrt{T}(\hat{\alpha}-\alpha_D)$ is bounded in probability, we obtain the well known result

(4.28)
$$\sqrt{T}(\hat{\alpha} - \alpha_D) = \sqrt{T} H_F^{-1} Q_F(\alpha_D) + o_p(1).$$

Similarly,

(4.29)
$$\sqrt{T}(\hat{\beta} - \beta_D) = \sqrt{T}H_G^{-1}Q_G(\beta_D) + o_p(1).$$

Combining (4.28) with the Taylor expansion

(4.30)
$$\sqrt{T}\beta_{\hat{\alpha}} = \sqrt{T}\beta_{\alpha_D} + \sqrt{T}B(\hat{\alpha} - \alpha_D) + o_p(1)$$

yields

(4.31)
$$\sqrt{T}(\beta_{\hat{\alpha}} - \beta_{\alpha_D}) = \sqrt{T}BH_F^{-1}Q_F(\alpha_D) + o_p(1).$$

In order to find a similar expression for $\sqrt{T}(\beta_{\hat{\alpha}}^* - \beta_{\alpha_D})$, we follow Gouriéroux–Monfort–Trognon [1983] and define the function

(4.32)
$$h(\alpha, \beta) = E_\alpha\left[Q_G(\beta)\right] = \frac{1}{T}\sum_{t=1}^{T} E_\alpha\left[Q_G^t(\beta)\right].$$

Its partial derivatives are given by

(4.33)
$$\frac{\partial h(\alpha, \beta)}{\partial \alpha'} = \frac{1}{T}\sum_{t=1}^{T} E_\alpha\left[Q_G^t(\beta)Q_F^{t\prime}(\alpha)\right]$$

and

(4.34)
$$\frac{\partial h(\alpha, \beta)}{\partial \beta'} = \frac{1}{T}\sum_{t=1}^{T} E_\alpha\left[H_G^t(\beta)\right].$$

Note that the definition of β_α^* implies that $h(\alpha, \beta_\alpha^*) = 0$ whenever β_α^* exists. Hence, expanding $\sqrt{T}h(\hat{\alpha}, \beta_{\hat{\alpha}}^*)$ around $\sqrt{T}h(\alpha_D, \beta_{\alpha_D})$ gives

(4.35)
$$0 = \sqrt{T}h(\alpha_D, \beta_{\alpha_D}) + \sqrt{T}\left[\frac{\partial h(\alpha, \beta)}{\partial \alpha'}\right]_{\alpha=\alpha_D,\beta=\beta_{\alpha_D}} (\hat{\alpha} - \alpha_D)$$
$$+ \sqrt{T}\left[\frac{\partial h(\alpha, \beta)}{\partial \beta'}\right]_{\alpha=\alpha_D,\beta=\beta_{\alpha_D}} (\beta_{\hat{\alpha}}^* - \beta_{\alpha_D}) + o_p(1).$$

Noting that

(4.36)
$$\left.\frac{\partial h(\alpha, \beta)}{\partial \alpha'}\right|_{\alpha=\alpha_D,\beta=\beta_{\alpha_D}} \xrightarrow[P]{a.s.} \bar{J}_{GF}$$

74

and

(4.37) $$\frac{\partial h(\alpha,\beta)}{\partial \beta'}\bigg|_{\alpha=\alpha_D,\beta=\beta_{\alpha_D}} \xrightarrow[P]{a.s.} -\bar{H}_G,$$

it follows from (4.35), (4.22) and (4.28) that

(4.38) $$\sqrt{T}(\beta_{\hat{\alpha}}^* - \beta_{\alpha_D}) = \sqrt{T}BH_F^{-1}Q_F(\alpha_D)$$
$$+ \sqrt{T}\bar{H}_G^{-1}E_{\alpha_D}\left[Q_G(\beta_{\alpha_D})\right] + o_p(1).$$

On the other hand, it follows from the multivariate central limit theorem that

(4.39) $$\sqrt{T}\begin{pmatrix} Q_F(\alpha_D) \\ E_{\alpha_D}\left[Q_G(\beta_{\alpha_D})\right] \end{pmatrix} \xrightarrow[\pi_D]{d} N\left(0, \begin{pmatrix} J_F & J_{F\bar{G}} \\ J_{\bar{G}F} & J_{\bar{G}} \end{pmatrix}\right).$$

The desired results follow from (4.31), (4.38) and (4.39). □

4.2. Asymptotic normality: further results

The following three theorems characterize the limit distributions of $\hat{\phi}$ and $\hat{\phi}^*$, of $\hat{q}$ and $\hat{q}^*$, and of $\hat{l}$ and $\hat{l}^*$, respectively, both under the encompassing hypothesis and under deviations from it. The limit distributions of $\hat{\phi}$, $\hat{\phi}^*$, $\hat{q}$ and $\hat{q}^*$ have been obtained by Gouriéroux–Monfort [1992] under the assumption that $\mathcal{F} \, \mathcal{E}_D^P \, \mathcal{G}$, though in the more general dynamic case.

Theorem 4.1. *Given Assumptions A1–A6, B1–B7 and N1–N5,*

(4.40) $$\sqrt{T}(\hat{\phi} - \phi) \xrightarrow[\pi_D]{d} N(0, V_\phi),$$

(4.41) $$\sqrt{T}(\hat{\phi}^* - \phi) \xrightarrow[\pi_D]{d} N(0, V_\phi^*),$$

where

(4.42) $V_\phi = H_G^{-1}\left(J_G - J_{GF}J_F^{-1}J_{FG}\right)H_G^{-1}$
$$+ H_G^{-1}\left(J_{GF}J_F^{-1} - H_G BH_F^{-1}\right)J_F\left(J_F^{-1}J_{FG} - H_F^{-1}B'H_G\right)H_G^{-1},$$

(4.43) $V_\phi^* = H_G^{-1}\left(J_G - J_{GF}J_F^{-1}J_{FG}\right)H_G^{-1}$
$$+ H_G^{-1}\left(J_{GF}J_F^{-1} - H_G BH_F^{-1}\right)J_F\left(J_F^{-1}J_{FG} - H_F^{-1}B'H_G\right)H_G^{-1}$$
$$+ \bar{H}_G^{-1}J_{\bar{G}}\bar{H}_G^{-1} + \bar{H}_G^{-1}\left(J_{\bar{G}F}H_F^{-1}B' - J_{\bar{G}G}H_G^{-1}\right)$$
$$+ \left(BH_F^{-1}J_{F\bar{G}} - H_G^{-1}J_{G\bar{G}}\right)\bar{H}_G^{-1}.$$

Proof. From (4.29), (4.31) and (4.38), it follows that

$$(4.44) \quad \sqrt{T}(\hat{\phi} - \phi) = \sqrt{T}(\hat{\beta} - \beta_D) - \sqrt{T}(\beta_{\hat{\alpha}} - \beta_{\alpha_D})$$
$$= -\sqrt{T}BH_F^{-1}Q_F(\alpha_D) + \sqrt{T}H_G^{-1}Q_G(\beta_D) + o_p(1),$$

$$(4.45) \quad \sqrt{T}(\hat{\phi}^* - \phi) = \sqrt{T}(\hat{\beta} - \beta_D) - \sqrt{T}(\beta_{\hat{\alpha}} - \beta_{\alpha_D})$$
$$= -\sqrt{T}BH_F^{-1}Q_F(\alpha_D) + \sqrt{T}H_G^{-1}Q_G(\beta_D)$$
$$- \sqrt{T}\tilde{H}_G^{-1}E_{\alpha_D}\left[Q_G(\beta_{\alpha_D})\right] + o_p(1).$$

Noting that, from the multivariate central limit theorem,

$$(4.46) \quad \sqrt{T}\begin{pmatrix} Q_F(\alpha_D) \\ Q_G(\beta_D) \\ E_{\alpha_D}\left[Q_G(\beta_{\alpha_D})\right] \end{pmatrix} \xrightarrow[\pi_D]{d} N\left(0, \begin{pmatrix} J_F & J_{FG} & J_{F\tilde{G}} \\ J_{GF} & J_G & J_{G\tilde{G}} \\ J_{\tilde{G}F} & J_{\tilde{G}G} & J_{\tilde{G}} \end{pmatrix}\right),$$

the results follow from (4.44)–(4.45) after rearranging. $\square$

Theorem 4.2. *Given Assumptions A1–A6, B1–B7 and N1–N5,*

$$(4.47) \quad \sqrt{T}(\hat{q} - q) \xrightarrow[\pi_D]{d} N(0, V_q),$$

$$(4.48) \quad \sqrt{T}(\hat{q}^* - q) \xrightarrow[\pi_D]{d} N(0, V_q^*),$$

where

$$(4.49) \quad V_q = \tilde{J}_G - \tilde{J}_{GF}J_F^{-1}\tilde{J}_{FG}$$
$$+ \left(\tilde{J}_{GF}J_F^{-1} - \tilde{H}_GBH_F^{-1}\right)J_F\left(J_F^{-1}\tilde{J}_{FG} - H_F^{-1}B'\tilde{H}_G\right),$$

$$(4.50) \quad V_q^* = \tilde{J}_G - \tilde{J}_{GF}J_F^{-1}\tilde{J}_{FG}$$
$$+ \left(\tilde{J}_{GF}J_F^{-1} - \tilde{H}_GBH_F^{-1}\right)J_F\left(J_F^{-1}\tilde{J}_{FG} - H_F^{-1}B'\tilde{H}_G\right)$$
$$+ \tilde{H}_G\tilde{H}_G^{-1}\tilde{J}_{\tilde{G}}\tilde{H}_G^{-1}\tilde{H}_G + \tilde{H}_G\tilde{H}_G^{-1}\left(\tilde{J}_{\tilde{G}F}H_F^{-1}B'\tilde{H}_G - \tilde{J}_{\tilde{G}G}\right)$$
$$+ \left(\tilde{H}_GBH_F^{-1}\tilde{J}_{F\tilde{G}} - \tilde{J}_{G\tilde{G}}\right)\tilde{H}_G^{-1}\tilde{H}_G.$$

Proof. Using (4.31) and (4.38) in the Taylor expansions

$$(4.51) \quad \sqrt{T}\hat{q} = \sqrt{T}Q_G(\beta_{\alpha_D}) + \sqrt{T}H_G(\beta_{\alpha_D})(\beta_{\hat{\alpha}} - \beta_{\alpha_D}) + o_p(1),$$

$$(4.52) \qquad \sqrt{T}\hat{q}^* = \sqrt{T}Q_G(\beta_{\alpha_D}) + \sqrt{T}H_G(\beta_{\alpha_D})(\beta_{\hat{\alpha}}^* - \beta_{\alpha_D}) + o_p(1),$$

we obtain

$$(4.53) \quad \sqrt{T}(\hat{q} - q) = -\sqrt{T}\tilde{H}_G B H_F^{-1} Q_F(\alpha_D) + \sqrt{T}\left(Q_G(\beta_{\alpha_D}) - q\right) + o_p(1),$$

$$(4.54) \quad \sqrt{T}(\hat{q}^* - q) = -\sqrt{T}\tilde{H}_G B H_F^{-1} Q_F(\alpha_D) + \sqrt{T}\left(Q_G(\beta_{\alpha_D}) - q\right)$$
$$- \sqrt{T}\tilde{H}_G \bar{H}_G^{-1} E_{\alpha_D}\left[Q_G(\beta_{\alpha_D})\right] + o_p(1).$$

From the central limit theorem, we have

$$(4.55) \quad \sqrt{T}\begin{pmatrix} Q_F(\alpha_D) \\ Q_G(\beta_{\alpha_D}) - q \\ E_{\alpha_D}\left[Q_G(\beta_{\alpha_D})\right] \end{pmatrix} \xrightarrow[\pi_D]{d} N\left(0, \begin{pmatrix} J_F & \tilde{J}_{FG} & J_{F\tilde{G}} \\ \tilde{J}_{GF} & \tilde{J}_G & \tilde{J}_{G\tilde{G}} \\ J_{\tilde{G}F} & \tilde{J}_{\tilde{G}G} & J_{\tilde{G}} \end{pmatrix}\right),$$

wherefrom the results are obtained in the usual way. $\qquad \square$

The distribution of a weighted sum of chi-squares arises as the limit distribution of $\hat{l}$ and of $\hat{l}^*$ under particular circumstances. This distribution is the object of the following definition.

Definition 4.1. *Let Ξ be an $n \times 1$ vector of independent χ_1^2 variates, and let λ be an $n \times 1$ vector of real numbers. Then the random variable $\lambda'\Xi$ is distributed as a weighted sum of chi-squares with parameter λ. This distribution is denoted by $M(\lambda)$.*

As is well known, the distribution $M(\lambda)$ arises as the distribution of quadratic forms in (asymptotically) normal variates with mean zero. More explicitly, if Z is (asymptotically) normally distributed with mean zero and $n \times n$ covariance matrix V, and W is an $n \times n$ matrix of real numbers, then $Z'WZ$ is (asymptotically) distributed as $M(\lambda(WV))$, where $\lambda(WV)$ denotes the $n \times 1$ vector of eigenvalues of WV. This property of quadratic forms will be used extensively in the sequel.

To ensure that the variance of $L_G^t(\beta_{\alpha_D}) - L_G^t(\beta_D)$ exists, we make the following assumption.

Assumption N6. *In some open set containing the line segment joining β_D and β_{α_D}, $|\log g(y|x;\beta)|^2$ is dominated by a π_D-integrable function independent of β.*

Assumption N6 ensures the existence of the following quantities:

$$(4.56) \qquad \omega_G = E_P E_D \left[L_G^t(\beta_{\alpha_D}) - L_G^t(\beta_D)\right]^2 - l^2,$$

(4.57) $$\omega_{FG} = E_P E_D \left[Q_F^t(\alpha_D) \left(L_G^t(\beta_{\alpha_D}) - L_G^t(\beta_D) \right) \right] = \omega_{GF}',$$

(4.58) $$\omega_{\bar{G}G} = E_P E_D \left[E_{\alpha_D} \left[Q_G^{t\prime}(\beta_{\alpha_D}) \right] \left(L_G^t(\beta_{\alpha_D}) - L_G^t(\beta_D) \right) \right] = \omega_{G\bar{G}}'.$$

The quantities defined in (4.56)–(4.58) are the π_D-almost sure limits of their sample analogues, denoted by $\hat{\omega}_G$, $\hat{\omega}_{FG}$ and $\hat{\omega}_{\bar{G}G}$, respectively, which are obtained by replacing $E_P E_D$ by $T^{-1} \sum_{t=1}^T$, E_P by $T^{-1} \sum_{t=1}^T$, q by $\hat{q}^*$, α_D by $\hat{\alpha}$, β_D by $\hat{\beta}$, and β_{α_D} by $\beta_{\hat{\alpha}}^*$, successively. The following theorem characterizes the limit distributions of $\hat{l}$ and $\hat{l}^*$.

Theorem 4.3. *Given Assumptions A1–A6, B1–B7 and N1–N5,*
(i) if $\mathcal{F} \, \mathcal{E}_D^P \, \mathcal{G}$*, then*

(4.59) $$-2T\hat{l} \xrightarrow[\pi_D]{d} M\left(\lambda(H_G V_\phi) \right),$$

(4.60) $$-2T\hat{l}^* \xrightarrow[\pi_D]{d} M\left(\lambda(H_G V_\phi^*) \right),$$

(ii) if $\mathcal{F} \, \mathcal{E}_D^P \, \mathcal{G}$ *and Assumption N6 holds, then*

(4.61) $$\sqrt{T}(\hat{l} - l) \xrightarrow[\pi_D]{d} N(0, V_l),$$

(4.62) $$\sqrt{T}(\hat{l}^* - l) \xrightarrow[\pi_D]{d} N(0, V_l^*),$$

where

(4.63) $$V_l = \omega_G + 2q' B H_F^{-1} \omega_{FG} + q' B H_F^{-1} J_F H_F^{-1} B' q,$$

(4.64) $$V_l^* = \omega_G + 2q' B H_F^{-1} \omega_{FG} + q' B H_F^{-1} J_F H_F^{-1} B' q$$
$$+ q' \bar{H}_G^{-1} J_{\bar{G}} \bar{H}_G^{-1} q + 2\left(\omega_{G\bar{G}} + q' B H_F^{-1} J_{F\bar{G}} \right) \bar{H}_G^{-1} q.$$

Proof. To prove (i), consider the Taylor expansions

(4.65) $$T L_G(\beta_{\hat{\alpha}}) = T L_G(\hat{\beta}) + T Q_G'(\hat{\beta})(\beta_{\hat{\alpha}} - \hat{\beta})$$
$$+ \frac{T}{2}(\beta_{\hat{\alpha}} - \hat{\beta})' H_G(\beta_*)(\beta_{\hat{\alpha}} - \hat{\beta}),$$

78

$$(4.66) \quad TL_G(\beta_{\hat\alpha}^*) = TL_G(\hat\beta) + TQ_G'(\hat\beta)(\beta_{\hat\alpha}^* - \hat\beta)$$
$$+ \frac{T}{2}(\beta_{\hat\alpha}^* - \hat\beta)' H_G(\beta_{**})(\beta_{\hat\alpha}^* - \hat\beta),$$

where β_* and β_{**} lie on the line segments joining $\beta_{\hat\alpha}$ and $\hat\beta$, and $\beta_{\hat\alpha}^*$ and $\hat\beta$, respectively. Since $\beta_* \xrightarrow[\pi_D]{a.s.} \beta_D$ and $\beta_{**} \xrightarrow[\pi_D]{a.s.} \beta_D$ when $\mathcal{F} \, \mathcal{E}_D^P \, \mathcal{G}$, it follows that

$$(4.67) \quad -2T\hat{l} = T\hat\phi' H_G \hat\phi + o_p(1),$$

$$(4.68) \quad -2T\hat{l}^* = T\hat\phi^{*\prime} H_G \hat\phi^* + o_p(1).$$

Part (i) follows. To prove part (ii), consider the Taylor expansions

$$(4.69) \quad \sqrt{T} L_G(\beta_{\alpha_D}) = \sqrt{T} L_G(\beta_{\hat\alpha}) - \sqrt{T} Q_G'(\beta_{\hat\alpha})(\beta_{\hat\alpha} - \beta_{\alpha_D}) + o_p(1),$$

$$(4.70) \quad \sqrt{T} L_G(\beta_{\alpha_D}) = \sqrt{T} L_G(\beta_{\hat\alpha}^*) - \sqrt{T} Q_G'(\beta_{\hat\alpha}^*)(\beta_{\hat\alpha}^* - \beta_{\alpha_D}) + o_p(1),$$

$$(4.71) \quad \sqrt{T} L_G(\beta_D) = \sqrt{T} L_G(\hat\beta) + o_p(1).$$

Since $\sqrt{T}(\beta_{\hat\alpha} - \beta_{\alpha_D})$ and $\sqrt{T}(\beta_{\hat\alpha}^* - \beta_{\alpha_D})$ are bounded in probability and $Q_G(\beta_{\hat\alpha}) \xrightarrow[\pi_D]{a.s} q$ and $Q_G(\beta_{\hat\alpha}^*) \xrightarrow[\pi_D]{a.s} q$, we obtain, using (4.31) and (4.38),

$$(4.72) \quad \sqrt{T}(\hat{l} - l) = \sqrt{T} q' B H_F^{-1} Q_F(\alpha_D)$$
$$+ \sqrt{T} \left(L_G(\beta_{\alpha_D}) - L_G(\beta_D) - l \right) + o_p(1),$$

$$(4.73) \quad \sqrt{T}(\hat{l}^* - l) = \sqrt{T} q' B H_F^{-1} Q_F(\alpha_D) + \sqrt{T} q' \bar{H}_G^{-1} E_{\alpha_D} \left[Q_G(\beta_{\alpha_D}) \right]$$
$$+ \sqrt{T} \left(L_G(\beta_{\alpha_D}) - L_G(\beta_D) - l \right) + o_p(1).$$

From the central limit theorem, we have

$$(4.74)$$
$$\sqrt{T} \begin{pmatrix} Q_F(\alpha_D) \\ E_{\alpha_D} \left[Q_G(\beta_{\alpha_D}) \right] \\ L_G(\beta_{\alpha_D}) - L_G(\beta_D) - l \end{pmatrix} \xrightarrow[\pi_D]{d} N \left(0, \begin{pmatrix} J_F & J_{F\bar G} & \omega_{FG} \\ J_{\bar G F} & J_{\bar G} & \omega_{\bar G G} \\ \omega_{GF} & \omega_{G\bar G} & \omega_G \end{pmatrix} \right).$$

Part (ii) follows from (4.72)–(4.74). $\qquad\Box$

Theorem 4.3 shows that the limit distributions of the modified likelihood ratio statistics depend on whether or not $\mathcal{F} \, \mathcal{E}_D^P \, \mathcal{G}$. This result is quite similar to the result obtained by Vuong [1989], who showed that the limit distribution of the usual likelihood ratio statistic depends on whether the stronger condition $F_{\alpha_D} = G_{\beta_D}$ holds or not. The following lemma supplements Lemma 3.1. It is similar to Lemma 4.1 in Vuong [1989].

Lemma 4.4. *Given Assumptions A1–A4, B1–B2, B5, B7 and N6,*

$$(4.75) \qquad \mathcal{F} \, \mathcal{E}_D^P \, \mathcal{G} \iff \omega_G = 0 \iff \hat{\omega}_G \xrightarrow[\pi_D]{a.s.} 0.$$

Proof. Clearly, $\omega_G = 0$ if and only if $L_G^t(\beta_{\alpha_D}) - L_G^t(\beta_D)$ is a constant, which then must be equal to zero, implying $l = 0$, i.e., $\mathcal{F} \, \mathcal{E}_D^P \, \mathcal{G}$. Noting that $\hat{\omega}_G \xrightarrow[\pi_D]{a.s.} \omega_G$, the proof is complete. $\qquad\square$

Theorems 4.1 to 4.3 characterize the limit distributions of $\hat{\phi}$, $\hat{\phi}^*$, $\hat{q}$, $\hat{q}^*$, $\hat{l}$ and $\hat{l}^*$ under general conditions. They constitute the main results of this chapter, and can be used to construct consistent tests of the encompassing hypothesis with correct asymptotic level. First, note that the covariance matrices characterizing the limit distributions can be consistently estimated from the data. Since the inverses of H_F, J_F and $\bar{H}_G$ appear in these covariance matrices, Assumptions N3 and N5 are crucial in this respect. Secondly, the critical points of a weighted sum of chi-squares distribution can be computed numerically. Hence, for any asymptotic level, a consistent likelihood ratio test follows immediately from Theorem 4.3, by considering $-2T\hat{l}^*$, or, when feasible, $-2T\hat{l}$. Consistent Wald and score tests, for any asymptotic level, follow from Theorems 4.1 and 4.2, respectively, by considering positive definite quadratic forms in $\hat{\phi}^*$ or $\hat{\phi}$ and $\hat{q}^*$ or $\hat{q}$, respectively. These quadratic forms, multiplied by T, all have a limit weighted sum of chi-squares distribution if $\mathcal{F} \, \mathcal{E}_D^P \, \mathcal{G}$, which is readily obtained from the above results. Otherwise, just as $-2T\hat{l}^*$ and $-2T\hat{l}$, they converge π_D-almost surely to $+\infty$, implying consistency. More generally, it is of interest to consider arbitrary quadratic forms in the Wald and score vectors, in particular using stochastic weighting matrices of incomplete rank. The limit distributions associated with this class of statistics are studied in Section 5 below. We note that the results obtained thus far, i.e., Theorems 4.1–4.3, are a prerequisite for a power analysis under fixed alternatives. This problem, however, is left for future work.

It has to be noted that, even if $\mathcal{F} \, \mathcal{E}_D^P \, \mathcal{G}$, the rank of the covariance matrices of the limit distributions of $\hat{\phi}^*$, $\hat{\phi}$, $\hat{q}^*$ and $\hat{q}$ depends on D and P, and hence is generally unknown. Similarly, if $\mathcal{F} \, \mathcal{E}_D^P \, \mathcal{G}$, $-2T\hat{l}$ and $-2T\hat{l}^*$ may or may not be asymptotically degenerate, depending on D and P. The possible non-constancy of the rank of the limiting covariance matrix under the null hypothesis is a general problem in statistical testing situations, and causes a difficulty for the construction of appropriate test statistics and for establishing their asymptotic distribution. For instance, it is a common practice to take, for example, $T\hat{\phi}'\hat{V}_\phi^+\hat{\phi}$ as a test statistic, where $\hat{V}_\phi^+$ is the Moore-Penrose inverse of $\hat{V}_\phi$, a consistent estimator of V_ϕ, and to claim that this statistic has a limiting chi-square distribution under the null hypothesis (i.e., $\mathcal{F} \mathcal{E}_D^P \mathcal{G}$). However, as pointed out by Andrews [1989], this claim is not true in general, but it is true under additional conditions. The problem crops up when the rank of $\hat{V}_\phi$ exceeds the rank of V_ϕ with probability bounded away from zero as $T \to \infty$. In this case, the reciprocal of the smallest eigenvalue of $\hat{V}_\phi$ becomes unboundedly large as $T \to \infty$ with positive probability, and hence $\hat{V}_\phi^+$ is inconsistent for V_ϕ^+. It is precisely the fact that the rank of V_ϕ is unknown which makes it difficult to design an estimator $\hat{V}_\phi$ whose rank converges in probability to the rank of V_ϕ. We shall return to this problem in Section 5 when studying the limit behaviour of quadratic forms.

The following corollary is obtained from Theorems 4.1–4.3 under the additional assumptions (i) $\mathcal{F} \mathcal{E}_D^P \mathcal{G}$, (ii) $D \in \mathcal{F}$, and (iii) $D \in \mathcal{F} \cap \mathcal{G}$, respectively. It follows by noting that (i) implies $\phi = q = 0$, $l = 0$, $\tilde{H}_G = H_G$, $\tilde{J}_G = J_G$, $\tilde{J}_{FG} = J_{FG} = \tilde{J}'_{GF}$ and $\tilde{J}_{G\tilde{G}} = J_{G\tilde{G}} = \tilde{J}'_{\tilde{G}G}$, (ii) implies $H_F = J_F$, $\tilde{H}_G = H_G$, $\tilde{J}_{FG} = J_{FG} = J'_{GF}$, $B = H_G^{-1}J_{GF}$ and $J_{G\tilde{G}} = J_{\tilde{G}} = J'_{\tilde{G}G}$, (iii) implies $H_G = J_G$, and that (iii) $\Rightarrow$ (ii) $\Rightarrow$ (i). The equality $H_F = J_F$ (resp. $H_G = J_G$), which holds whenever $D \in \mathcal{F}$ (resp. $D \in \mathcal{G}$), is known as the information matrix equivalence.

Corollary 4.1. *Let Assumptions A1–A6, B1–B7 and N1–N5 hold.*
(i) If $\mathcal{F} \mathcal{E}_D^P \mathcal{G}$, then

$$(4.76) \qquad \sqrt{T}\hat{\phi} \xrightarrow[\pi_D]{d} N(0, V_\phi),$$

$$(4.77) \qquad \sqrt{T}\hat{\phi}^* \xrightarrow[\pi_D]{d} N(0, V_\phi^*),$$

$$(4.78) \qquad \sqrt{T}\hat{q} \xrightarrow[\pi_D]{d} N(0, H_G V_\phi H_G),$$

$$(4.79) \qquad \sqrt{T}\hat{q}^* \xrightarrow[\pi_D]{d} N(0, H_G V_\phi^* H_G).$$

(ii) If $D \in \mathcal{F}$, then

$$(4.80) \qquad \sqrt{T}\hat{\phi} \xrightarrow[\pi_D]{d} N\big(0, H_G^{-1}(J_G - J_{GF}J_F^{-1}J_{FG})H_G^{-1}\big),$$

$$(4.81) \quad \sqrt{T}\hat{\phi}^* \xrightarrow[\pi_D]{d} N\big(0, H_G^{-1}(J_G - J_{GF}J_F^{-1}J_{FG})H_G^{-1}$$
$$- H_G^{-1}(J_{\tilde{G}} - J_{\tilde{G}F}J_F^{-1}J_{FG} - J_{GF}J_F^{-1}J_{F\tilde{G}})H_G^{-1}\big),$$

$$(4.82) \quad \sqrt{T}\hat{q} \xrightarrow[\pi_D]{d} N(0, J_G - J_{GF}J_F^{-1}J_{FG}),$$

$$(4.83) \quad \sqrt{T}\hat{q}^* \xrightarrow[\pi_D]{d} N\big(0, J_G - J_{GF}J_F^{-1}J_{FG}$$
$$- (J_{\tilde{G}} - J_{\tilde{G}F}J_F^{-1}J_{FG} - J_{GF}J_F^{-1}J_{F\tilde{G}})\big),$$

$$(4.84) \quad -2T\hat{l} \xrightarrow[\pi_D]{d} M\big(\lambda(J_G H_G^{-1} - J_{GF}J_F^{-1}J_{FG}H_G^{-1})\big),$$

$$(4.85) \quad -2T\hat{l}^* \xrightarrow[\pi_D]{d} M\big(\lambda\big(J_G H_G^{-1} - J_{GF}J_F^{-1}J_{FG}H_G^{-1}$$
$$- (J_{\tilde{G}} - J_{\tilde{G}F}J_F^{-1}J_{FG} - J_{GF}J_F^{-1}J_{F\tilde{G}})H_G^{-1}\big)\big).$$

(iii) If $D \in \mathcal{F} \cap \mathcal{G}$, then

$$(4.86) \quad \sqrt{T}\hat{\phi} \xrightarrow[\pi_D]{d} N(0, J_G^{-1} - J_G^{-1}J_{GF}J_F^{-1}J_{FG}J_G^{-1}),$$

$$(4.87) \quad \sqrt{T}\hat{\phi}^* \xrightarrow[\pi_D]{d} N\big(0, J_G^{-1} - J_G^{-1}J_{GF}J_F^{-1}J_{FG}J_G^{-1}$$
$$- J_G^{-1}(J_{\tilde{G}} - J_{\tilde{G}F}J_F^{-1}J_{FG} - J_{GF}J_F^{-1}J_{F\tilde{G}})J_G^{-1}\big),$$

$$(4.88) \quad \sqrt{T}\hat{q} \xrightarrow[\pi_D]{d} N(0, J_G - J_{GF}J_F^{-1}J_{FG}),$$

$$(4.89) \quad \sqrt{T}\hat{q}^* \xrightarrow[\pi_D]{d} N\big(0, J_G - J_{GF}J_F^{-1}J_{FG}$$
$$- (J_{\tilde{G}} - J_{\tilde{G}F}J_F^{-1}J_{FG} - J_{GF}J_F^{-1}J_{F\tilde{G}})\big),$$

$$(4.90) \quad -2T\hat{l} \xrightarrow[\pi_D]{d} M\big(\lambda(I - J_{GF}J_F^{-1}J_{FG}J_G^{-1})\big),$$

$$(4.91) \quad -2T\hat{l}^* \xrightarrow[\pi_D]{d} M\big(\lambda\big(I - J_{GF}J_F^{-1}J_{FG}J_G^{-1}$$
$$- (J_{\tilde{G}} - J_{\tilde{G}F}J_F^{-1}J_{FG} - J_{GF}J_F^{-1}J_{F\tilde{G}})J_G^{-1}\big)\big).$$

Part (i) has already been obtained by Gouriéroux–Monfort [1995], and (4.80)–(4.83) by Gouriéroux–Monfort–Trognon [1983].

4.3. Asymptotic equivalences

A final question of interest here concerns asymptotic equivalences between $\hat{\phi}$, $\hat{q}$ and $\hat{l}$, and between $\hat{\phi}^*$, $\hat{q}^*$ and $\hat{l}^*$. The following theorem establishes such equivalences, given that $\mathcal{F} \; \mathcal{E}_D^P \; \mathcal{G}$, thereby generalizing some well known results. In the sequel, $\xrightarrow[\pi_D]{p}$ denotes convergence in probability under π_D as $T \to \infty$.

Theorem 4.4. *Given Assumptions A1–A6, B1–B7 and N1–N5, if $\mathcal{F} \; \mathcal{E}_D^P \; \mathcal{G}$, then*

$$(4.92) \qquad \sqrt{T}(\hat{\phi} - H_G^{-1}\hat{q}) \xrightarrow[\pi_D]{p} 0,$$

$$(4.93) \qquad \sqrt{T}(\hat{\phi}^* - H_G^{-1}\hat{q}^*) \xrightarrow[\pi_D]{p} 0,$$

$$(4.94) \qquad T(\hat{\phi}'H_G\hat{\phi} + 2\hat{l}) \xrightarrow[\pi_D]{p} 0,$$

$$(4.95) \qquad T(\hat{\phi}^{*\prime}H_G\hat{\phi}^* + 2\hat{l}^*) \xrightarrow[\pi_D]{p} 0,$$

$$(4.96) \qquad T(\hat{q}'H_G^{-1}\hat{q} + 2\hat{l}) \xrightarrow[\pi_D]{p} 0,$$

$$(4.97) \qquad T(\hat{q}^{*\prime}H_G^{-1}\hat{q}^* + 2\hat{l}^*) \xrightarrow[\pi_D]{p} 0.$$

Proof. Assuming $\mathcal{F} \; \mathcal{E}_D^P \; \mathcal{G}$, it follows from (4.44)–(4.45) that

$$(4.98) \qquad \sqrt{T}\hat{\phi} = -\sqrt{T}BH_F^{-1}Q_F(\alpha_D) + \sqrt{T}H_G^{-1}Q_G(\beta_D) + o_p(1),$$

$$(4.99) \qquad \sqrt{T}\hat{\phi}^* = -\sqrt{T}BH_F^{-1}Q_F(\alpha_D) + \sqrt{T}H_G^{-1}Q_G(\beta_D)$$
$$- \sqrt{T}\bar{H}_G^{-1}E_{\alpha_D}[Q_G(\beta_D)] + o_p(1).$$

On the other hand, it follows from (4.53)–(4.54) that

$$(4.100) \qquad \sqrt{T}\hat{q} = -\sqrt{T}H_GBH_F^{-1}Q_F(\alpha_D) + \sqrt{T}Q_G(\beta_D) + o_p(1),$$

$$(4.101) \qquad \sqrt{T}\hat{q}^* = -\sqrt{T}H_GBH_F^{-1}Q_F(\alpha_D) + \sqrt{T}Q_G(\beta_D)$$
$$- \sqrt{T}H_G\bar{H}_G^{-1}E_{\alpha_D}[Q_G(\beta_D)] + o_p(1).$$

Hence, (4.92)–(4.93) follow. The results given by (4.94)–(4.95) have already been obtained in (4.67)–(4.68), and (4.96)–(4.97) are immediate consequences of (4.92)–(4.95). □

In general, no similar asymptotic equivalences between $\sqrt{T}(\hat{\phi} - \phi)$, $\sqrt{T}(\hat{q} - q)$ and $\sqrt{T}(\hat{l} - l)$, or between $\sqrt{T}(\hat{\phi}^* - \phi)$, $\sqrt{T}(\hat{q}^* - q)$ and $\sqrt{T}(\hat{l}^* - l)$ exist when $\mathcal{F} \not\overset{P}{\underset{D}{\mathcal{L}}} \mathcal{G}$. This is a consequence of the fact that, in contrast to, for example, (4.98)–(4.101), the first-order asymptotic representations of these statistics are not perfectly correlated. In fact, this suggests that one might gain in power by considering tests based on $\hat{\phi}$, $\hat{q}$ and $\hat{l}$ (resp. $\hat{\phi}^*$, $\hat{q}^*$ and $\hat{l}^*$) simultaneously instead of tests based on only one of $\hat{\phi}$, $\hat{q}$ and $\hat{l}$ (resp. $\hat{\phi}^*$, $\hat{q}^*$ and $\hat{l}^*$). We will not pursue this idea any further here. It would require the joint limit distribution of $\sqrt{T}(\hat{\phi} - \phi)$, $\sqrt{T}(\hat{q} - q)$ and $\sqrt{T}(\hat{l} - l)$, (resp. $\sqrt{T}(\hat{\phi}^* - \phi)$, $\sqrt{T}(\hat{q}^* - q)$ and $\sqrt{T}(\hat{l}^* - l)$) which may be found along the lines of the proofs of Theorems 4.1–4.3. Finally, note the similarities and differences between Theorem 4.4 and Lemma 4.1, which relates the non-random quantities ϕ, q and l.

5. Limit distributions of quadratic forms

Wald and score tests for the general purpose of testing encompassing hypotheses are based on quadratic forms in $\hat{\phi}$ or $\hat{\phi}^*$, and $\hat{q}$ or $\hat{q}^*$, respectively. In the previous section, we derived the limit distributions of $\hat{\phi}$, $\hat{\phi}^*$, $\hat{q}$ and $\hat{q}^*$ under general conditions. What remains is to do the same for quadratic forms in these statistics. Given the asymptotic normality of all these statistics, we need to study the asymptotic behaviour of quadratic forms in asymptotically normal variates.

We denote the vectors $\hat{\phi}$, $\hat{\phi}^*$, $\hat{q}$ and $\hat{q}^*$ generically by $\hat{\xi}$, the covariance matrices of their limit distributions by V_ξ, and the weighting matrices in their quadratic forms by W_ξ, $\hat{W}_\xi$ or $\hat{V}_\xi^+$. The following lemma gives a result for a fixed weighting matrix. $O_p(T^a)$ denotes a quantity that is at most of order T^a in π_D-probability.

Lemma 5.1. *Let*

$$(5.1) \qquad \sqrt{T}(\hat{\xi} - \xi) \xrightarrow[\pi_D]{d} N(0, V_\xi),$$

where ξ is $n \times 1$ and non-stochastic. Let W_ξ be an $n \times n$ non-stochastic symmetric positive semi-definite matrix.

(i) If $W_\xi \xi = 0$, then

$$(5.2) \qquad T\hat{\xi}' W_\xi \hat{\xi} \xrightarrow[\pi_D]{d} M\big(\lambda(W_\xi V_\xi)\big).$$

(ii) If $V_\xi W_\xi \xi \neq 0$, then

$$(5.3) \qquad \sqrt{T}(\hat{\xi}' W_\xi \hat{\xi} - \xi' W_\xi \xi) \xrightarrow[\pi_D]{d} N(0, 4\xi' W_\xi V_\xi W_\xi \xi).$$

(iii) $T\hat{\xi}' W_\xi \hat{\xi} = O_p(1)$ if and only if $W_\xi \xi = 0$.
(iv) $T\hat{\xi}' W_\xi \hat{\xi} \xrightarrow[\pi_D]{p} +\infty$ if and only if $W_\xi \xi \neq 0$.

Proof. Part (i) follows immediately from (5.1) by noting that $T\hat{\xi}' W_\xi \hat{\xi} = T(\hat{\xi} - \xi)' W_\xi (\hat{\xi} - \xi)$ if $W_\xi \xi = 0$. To prove part (ii), we write

$$
\begin{aligned}
\sqrt{T}(\hat{\xi}' W_\xi \hat{\xi} - \xi' W_\xi \xi) &= \sqrt{T}(\hat{\xi} - \xi)' W_\xi (\hat{\xi} - \xi) + 2\sqrt{T}\xi' W_\xi (\hat{\xi} - \xi) \\
(5.4) &= 2\sqrt{T}\xi' W_\xi (\hat{\xi} - \xi) + O_p(T^{-1/2}).
\end{aligned}
$$

Part (ii) follows. To prove parts (iii) and (iv), rewrite (5.4) as

$$(5.5) \qquad T\hat{\xi}' W_\xi \hat{\xi} = T\xi' W_\xi \xi + 2T\xi' W_\xi (\hat{\xi} - \xi) + O_p(1),$$

and observe that the first term on the RHS of (5.5) dominates the second one in π_D-probability as $T \to \infty$, unless $W_\xi \xi = 0$. This establishes parts (iii) and (iv). □

It is to be noted that the lemma does not completely characterize the first-order asymptotic behaviour of $T\hat{\xi}' W_\xi \hat{\xi}$. In particular, the ·case $V_\xi W_\xi \xi = 0$ and $W_\xi \xi \neq 0$ is not covered. Filling up this gap would require an asymptotic representation of $\hat{\xi}$ up to $o_p(T^{-1})$ instead of $o_p(T^{-1/2})$. As for $\hat{\phi}$, $\hat{\phi}^*$, $\hat{q}$ and $\hat{q}^*$, we would need stronger results than those of the previous section, by using higher order expansions than those considered. This is beyond the scope of the present study.

Although sometimes useful, the result of Lemma 5.1 is seriously limited by the non-stochastic nature of the weighting matrix W_ξ. Most test statistics use a random weighting matrix. The following generalization may then be useful. The notation $\otimes$ refers to the Kronecker product.

Lemma 5.2. *Let*

$$(5.6) \qquad \sqrt{T}\begin{pmatrix} \hat{\xi} - \xi \\ \mathrm{vec}(\hat{W}_\xi - W_\xi) \end{pmatrix} \xrightarrow[\pi_D]{d} N(0, V_{\xi*}),$$

85

where ξ is $n \times 1$ and non-stochastic and W_ξ is $n \times n$, non-stochastic, symmetric and positive semi-definite. Let also $\hat{W}_\xi$ be symmetric. Let V_ξ be the upper left $n \times n$ block of $V_{\xi*}$, and define

$$(5.7) \qquad \xi_* = \begin{pmatrix} 2W_\xi\xi \\ \xi \otimes \xi \end{pmatrix}.$$

(i) If $\xi = 0$, then

$$(5.8) \qquad T\hat{\xi}'\hat{W}_\xi\hat{\xi} \xrightarrow[\pi_D]{d} M\big(\lambda(W_\xi V_\xi)\big).$$

(ii) If $V_{\xi*}\xi_* \neq 0$, then

$$(5.9) \qquad \sqrt{T}(\hat{\xi}'\hat{W}_\xi\hat{\xi} - \xi'W_\xi\xi) \xrightarrow[\pi_D]{d} N(0, \xi_*'V_{\xi*}\xi_*).$$

(iii) $T\hat{\xi}'\hat{W}_\xi\hat{\xi} = O_p(1)$ if and only if $T\xi'\hat{W}_\xi\xi = O_p(1)$.
(iv) $T\hat{\xi}'\hat{W}_\xi\hat{\xi} \xrightarrow[\pi_D]{p} +\infty$ if and only if $T\xi'\hat{W}_\xi\xi \neq O_p(1)$.

Proof. Part (i) follows immediately by noting that $T\hat{\xi}'\hat{W}_\xi\hat{\xi} = T\hat{\xi}'W_\xi\hat{\xi} + o_p(1)$ if $\xi = 0$. To prove part (ii), we write

$$\sqrt{T}(\hat{\xi}'\hat{W}_\xi\hat{\xi} - \xi'W_\xi\xi) = \sqrt{T}(\hat{\xi} - \xi)'(\hat{W}_\xi - W_\xi)(\hat{\xi} - \xi) + \sqrt{T}(\hat{\xi} - \xi)'W_\xi(\hat{\xi} - \xi)$$
$$+ 2\sqrt{T}\xi'(\hat{W}_\xi - W_\xi)(\hat{\xi} - \xi) + \sqrt{T}\xi'(\hat{W}_\xi - W_\xi)\xi$$
$$+ 2\sqrt{T}\xi'W_\xi(\hat{\xi} - \xi)$$
$$= \sqrt{T}\xi'(\hat{W}_\xi - W_\xi)\xi + 2\sqrt{T}\xi'W_\xi(\hat{\xi} - \xi) + O_p(T^{-1/2})$$
$$(5.10) \qquad = \sqrt{T}\xi_*' \begin{pmatrix} \hat{\xi} - \xi \\ \mathrm{vec}(\hat{W}_\xi - W_\xi) \end{pmatrix} + O_p(T^{-1/2}).$$

Part (ii) follows. To prove parts (iii) and (iv), rewrite (5.10) as

$$(5.11) \qquad T\hat{\xi}'\hat{W}_\xi\hat{\xi} = T\xi'\hat{W}_\xi\xi + 2T\xi'W_\xi(\hat{\xi} - \xi) + O_p(1).$$

Again, the first term on the RHS of (5.11) dominates the second one, unless perhaps $W_\xi\xi = 0$, whence the result. $\qquad \square$

Again, we note that the case $V_{\xi*}\xi_* = 0$ and $\xi \neq 0$ is not covered. Furthermore, the application of Lemma 5.2 requires that $\hat{W}_\xi$ be a consistent estimator of W_ξ, as indicated by (5.6). In many cases, however, one wants to use as a weighting matrix the Moore-Penrose inverse of a symmetric consistent estimator of V_ξ, say $\hat{V}_\xi$, which is asymptotically normally distributed. When V_ξ is non-singular, a Taylor expansion yields

$$(5.12) \quad \sqrt{T}\,\text{vec}(\hat{V}_\xi^{-1} - V_\xi^{-1}) = -\sqrt{T}(V_\xi^{-1} \otimes V_\xi^{-1})\,\text{vec}(\hat{V}_\xi - V_\xi) + o_p(1),$$

and Lemma 5.2 can be applied with $W_\xi = V_\xi^{-1}$ and $\hat{W}_\xi = \hat{V}_\xi^{-1}$ to yield the limit behaviour of $\hat{\xi}'\hat{V}_\xi^{-1}\hat{\xi}$. For singular V_ξ, the inverses in (5.12) cannot be simply replaced by Moore-Penrose inverses, since the discontinuity of the Moore-Penrose inverse of singular matrices precludes the use of a Taylor expansion. In particular, if $\hat{V}_\xi$ is consistent for V_ξ, then $\hat{V}_\xi^+$ is consistent for V_ξ^+ if and only if the probability that $\text{rank}(\hat{V}_\xi) = \text{rank}(V_\xi)$ converges to one as $T \to \infty$—see Andrews [1989]. There are, however, ample situations where V_ξ is singular and $\hat{V}_\xi$, though being consistent for V_ξ, has full rank with probability one for all T. Hence, without further conditions, no general statements about the limit behaviour of $\hat{\xi}'\hat{V}_\xi^+\hat{\xi}$ can be derived from Lemma 5.2. In particular, it does not follow from part (i) that, if $\xi = 0$, $T\hat{\xi}'\hat{V}_\xi^+\hat{\xi}$ has a limiting chi-square distribution, unless some additional condition is satisfied. Andrews [1989] established a necessary and sufficient condition for this result to hold. We repeat this condition below. Also, part (ii) of Lemma 5.2 cannot be applied to yield the limit behaviour of $\sqrt{T}(\hat{\xi}'\hat{V}_\xi^+\hat{\xi} - \xi'V_\xi^+\xi)$. Along the same lines as Andrews [1989], we derive a similar condition that is sufficient to establish the asymptotic normality of $\sqrt{T}(\hat{\xi}'\hat{V}_\xi^+\hat{\xi} - \xi'V_\xi^+\xi)$.

Let $r = \text{rank}(V_\xi)$ and $\hat{r} = \text{rank}(\hat{V}_\xi)$. As indicated above, the problems associated with the quadratic form $\hat{\xi}'\hat{V}_\xi^+\hat{\xi}$ arise when $\text{Pr}_{\pi_D}[\hat{r} = r]$ does not converge to one as $T \to \infty$. Whenever $\hat{r} \geq r$, we can write $\hat{V}_\xi$ as

$$(5.13) \quad \hat{V}_\xi = \begin{cases} \tilde{P}\tilde{\Delta}\tilde{P}', & \text{if } \hat{r} = r, \\ \tilde{P}\tilde{\Delta}\tilde{P}' + \bar{P}\bar{\Delta}\bar{P}', & \text{if } \hat{r} > r, \end{cases}$$

where $\tilde{\Delta}$ and $\bar{\Delta}$ are $r \times r$ and $(\hat{r} - r) \times (\hat{r} - r)$ positive definite diagonal matrices, respectively, whose diagonal elements are the r largest and the $\hat{r} - r$ smallest non-zero eigenvalues of $\hat{V}_\xi$, respectively, and where $\tilde{P}$ and $\bar{P}$ are $n \times r$ and $n \times (\hat{r} - r)$ semi-orthogonal matrices, respectively (i.e., $\tilde{P}'\tilde{P} = I_r$ and $\bar{P}'\bar{P} = I_{\hat{r}-r}$), with $\tilde{P}'\bar{P} = 0$. The columns of $\tilde{P}$ and $\bar{P}$ are sets of eigenvectors of $\hat{V}_\xi$, associated with the r largest and the $\hat{r} - r$

smallest non-zero eigenvalues of $\hat{V}_\xi$, respectively. Let $\tilde{V}_\xi = \tilde{P}\tilde{\Delta}\tilde{P}'$ and let $\tilde{V}_\xi = 0$ when $\hat{r} = r$ and $\bar{V}_\xi = \bar{P}\bar{\Delta}\bar{P}'$ when $\hat{r} > r$. Finally, let

$$(5.14) \qquad \bar{b} = \begin{cases} 0, & \text{if } \hat{r} \leq r, \\ \hat{\xi}'\bar{V}_\xi^+\hat{\xi}, & \text{if } \hat{r} > r. \end{cases}$$

As the lemma below shows, the limit behaviour of $\bar{b}$ is crucial in determining the limit behaviour of $\hat{\xi}'\hat{V}_\xi^+\hat{\xi}$. Part (i) has already been proved by Andrews [1989, Theorem 1, part (a)].

Lemma 5.3. *Let*

$$(5.15) \qquad \sqrt{T}\begin{pmatrix} \hat{\xi} - \xi \\ \text{vec}(\hat{V}_\xi - V_\xi) \end{pmatrix} \xrightarrow[\pi_D]{d} N(0, V_{\xi*}),$$

where ξ is $n \times 1$ and non-stochastic and V_ξ is the upper left $n \times n$ block of $V_{\xi*}$. Let $\hat{V}_\xi$ be symmetric. Define

$$(5.16) \qquad \xi_+ = \begin{pmatrix} 2V_\xi^+\xi \\ -(V_\xi^+\xi) \otimes (V_\xi^+\xi) \end{pmatrix}.$$

Let $r = \text{rank}(V_\xi)$ and $\hat{r} = \text{rank}(\hat{V}_\xi)$. Write $\hat{V}_\xi$ as in (5.13) and $\bar{b}$ as in (5.14).
(i) If $\xi = 0$ and $T\bar{b} = o_p(1)$, then

$$(5.17) \qquad T\hat{\xi}'\hat{V}_\xi^+\hat{\xi} \xrightarrow[\pi_D]{d} \chi_r^2.$$

(ii) If $V_{\xi*}\xi_+ \neq 0$ and $\sqrt{T}\bar{b} = o_p(1)$, then

$$(5.18) \qquad \sqrt{T}(\hat{\xi}'\hat{V}_\xi^+\hat{\xi} - \xi'V_\xi^+\xi) \xrightarrow[\pi_D]{d} N(0, \xi_+'V_{\xi*}\xi_+).$$

(iii) $T\hat{\xi}'\hat{V}_\xi^+\hat{\xi} = O_p(1)$ if and only if $V_\xi\xi = 0$ and $T\bar{b} = O_p(1)$.
(iv) $T\hat{\xi}'\hat{V}_\xi^+\hat{\xi} \xrightarrow[\pi_D]{p} +\infty$ if and only if $V_\xi\xi \neq 0$ or $T\bar{b} \neq O_p(1)$.

Proof. With probability that goes to one, $\hat{r} \geq r$ and $\hat{V}_\xi$ can be written in a form given by (5.13). Although $\tilde{V}_\xi$ and $\bar{V}_\xi$ are not defined uniquely when $\hat{r} > r$ and the r-th and the $r + 1$-th largest eigenvalues of $\hat{V}_\xi$ are identical, this occurs with probability that goes to zero and hence $\tilde{V}_\xi$ and $\bar{V}_\xi$ are defined uniquely with probability that goes to one—see Andrews [1989]. Since $\tilde{V}_\xi$ is consistent for V_ξ and the rank of $\tilde{V}_\xi$ converges in probability to the rank of V_ξ, it follows that $\tilde{V}_\xi^+$ is consistent for V_ξ^+. Note furthermore that $\hat{V}_\xi^+ = \tilde{V}_\xi^+ + \bar{V}_\xi^+$. Hence, if $\xi = 0$,

$$T\hat{\xi}'\hat{V}_\xi^+\hat{\xi} = T\hat{\xi}'\tilde{V}_\xi^+\hat{\xi} + T\hat{\xi}'\bar{V}_\xi^+\hat{\xi}$$
$$(5.19) \qquad\qquad = T\hat{\xi}'\tilde{V}_\xi^+\hat{\xi} + T\bar{b} + o_p(1),$$

and part (i) follows. To prove part (ii), note that

$$(5.20) \qquad \sqrt{T}(\hat{\xi}'\hat{V}_\xi^+\hat{\xi} - \xi'V_\xi^+\xi) = \sqrt{T}(\hat{\xi}'\tilde{V}_\xi^+\hat{\xi} - \xi'V_\xi^+\xi) + \sqrt{T}\bar{b}.$$

Similarly to (5.10), the first term on the RHS of (5.20) can be written as

$$(5.21) \quad \sqrt{T}(\hat{\xi}'\tilde{V}_\xi^+\hat{\xi} - \xi'V_\xi^+\xi) = \sqrt{T}(2\xi'V_\xi^+ \quad \xi'\otimes\xi')\begin{pmatrix} \hat{\xi}-\xi \\ \mathrm{vec}(\tilde{V}_\xi^+ - V_\xi^+) \end{pmatrix}$$
$$+ O_p(T^{-1/2}).$$

Since $\tilde{V}_\xi$ has constant rank whenever it is defined, a Taylor expansion yields

$$\sqrt{T}\,\mathrm{vec}(\tilde{V}_\xi^+ - V_\xi^+) = -\sqrt{T}(V_\xi^+\otimes V_\xi^+)\,\mathrm{vec}(\tilde{V}_\xi - V_\xi) + O_p(T^{-1/2})$$
$$(5.22) \qquad\qquad = -\sqrt{T}(V_\xi^+\otimes V_\xi^+)\,\mathrm{vec}(\hat{V}_\xi - V_\xi)$$
$$+ \sqrt{T}(V_\xi^+\otimes V_\xi^+)\,\mathrm{vec}\,\bar{V}_\xi + O_p(T^{-1/2}).$$

Substituting (5.22) into (5.21) gives

$$(5.23) \qquad \sqrt{T}(\hat{\xi}'\tilde{V}_\xi^+\hat{\xi} - \xi'V_\xi^+\xi) = \sqrt{T}\xi'_+\begin{pmatrix} \hat{\xi}-\xi \\ \mathrm{vec}(\hat{V}_\xi - V_\xi) \end{pmatrix}$$
$$+ \sqrt{T}\xi'V_\xi^+\bar{V}_\xi V_\xi^+\xi + O_p(T^{-1/2}).$$

Noting that, since $\tilde{P}'\bar{P} = 0$,

$$(5.24) \qquad \sqrt{T}\xi'V_\xi^+\bar{V}_\xi V_\xi^+\xi = \sqrt{T}\xi'\tilde{V}_\xi^+\bar{V}_\xi\tilde{V}_\xi^+\xi + O_p(T^{-1/2}) = O_p(T^{-1/2}),$$

it follows from (5.15), (5.20) and (5.23) that

$$(5.25) \quad \sqrt{T}(\hat{\xi}'\hat{V}_\xi^+\hat{\xi} - \xi'V_\xi^+\xi) = \sqrt{T}\xi'_+\begin{pmatrix} \hat{\xi}-\xi \\ \mathrm{vec}(\hat{V}_\xi - V_\xi) \end{pmatrix} + \sqrt{T}\bar{b} + O_p(T^{-1/2}),$$

wherefrom part (ii) obtains. To prove parts (iii) and (iv), rewrite (5.25) as

$$(5.26) \qquad T\hat{\xi}'\hat{V}_\xi^+\hat{\xi} = T\xi'V_\xi^+\xi + 2T\xi'V_\xi^+(\hat{\xi}-\xi)$$
$$- T\xi'V_\xi^+(\hat{V}_\xi - V_\xi)V_\xi^+\xi + T\bar{b} + O_p(1).$$

Noting that $V_\xi^+\xi = 0$ if and only if $V_\xi\xi = 0$, the results follow in the usual way. $\qquad\square$

It is worth pointing out that, since $\bar{b} \geq 0$, the conditions $T\bar{b} = o_p(1)$ and $\sqrt{T}\bar{b} = o_p(1)$ are in fact necessary and sufficient for (5.17) and (5.18), respectively, to hold—see Lemma 2 in Andrews [1989].

Part (i) of the preceding lemmas will be useful for determining the limit distribution of a large class of encompassing test statistics under the encompassing hypothesis. Part (iv) shall be used to characterize the implicit null associated with the encompassing tests.

6. Encompassing tests

We are now ready to construct a large class of encompassing tests. Recall from Lemma 3.1 that, under suitable regularity conditions, the encompassing hypothesis is equivalent to each of the statistics $\hat{\phi}$, $\hat{\phi}^*$, $\hat{q}$, $\hat{q}^*$, $\hat{l}$ and $\hat{l}^*$ converging π_D-almost surely to zero. In Section 4, the limit distributions of these statistics were obtained under general conditions. As for $\hat{l}$ and $\hat{l}^*$, we showed that its asymptotic behaviour depends critically on whether $\mathcal{F}$ encompasses $\mathcal{G}$ or not. The other statistics, being vectors in general, were shown to be asymptotically normally distributed. Hence, by taking appropriate quadratic forms in these statistics, we can apply the results of Section 5 to arrive at similar conclusions as for $\hat{l}$ and $\hat{l}^*$. That is, the limit behaviour of appropriate quadratic forms in $\hat{\phi}$, $\hat{\phi}^*$, $\hat{q}$ and $\hat{q}^*$ depends critically on whether $\mathcal{F}$ (completely or incompletely) encompasses $\mathcal{G}$ or not. The possibility of choosing the weighting matrix used in these quadratic forms gives rise to a large class of statistics and associated tests, with different power properties in general. For each of these statistics, we define an encompassing test with asymptotic level ϵ between 0 and 1, by deriving the associated critical region. No power analysis will be carried out, except that we characterize the implicit null associated with each test.

We adopt the following generic notation. A distinction is drawn between likelihood ratio statistics and tests on the one hand, and other statistics and tests on the other hand. The statistics $\hat{\phi}$, $\hat{\phi}^*$, $\hat{q}$ and $\hat{q}^*$ are generically denoted by $\hat{\xi}$, their π_D-almost sure limits by ξ, the matrices V_ϕ, V_ϕ^*, V_q and V_q^* by V_ξ, and their estimators by $\hat{V}_\xi$. Furthermore, let $\hat{\zeta}$ denote $\hat{l}$ and $\hat{l}^*$, ζ their π_D-almost sure limits, V_ζ the matrices V_ϕ and V_ϕ^*, and $\hat{V}_\zeta$ their estimators.

It may happen that $V_\xi = 0$ (and similarly $V_\zeta = 0$). For instance, if $\hat{\alpha} = \hat{\beta}$ and $\beta_\alpha = \alpha$ for all $\alpha \in \Omega_\mathcal{F}$, then $\mathcal{F} \, \mathcal{E}_D^P \, \mathcal{G}$ and $V_\xi = 0$ for all ξ. Obviously, it may also be the case that $\mathcal{F} \, \mathcal{E}_D^P \, \mathcal{G}$ and $V_\xi \neq 0$. As yet, we do not know whether the reverse can occur, i.e., $\mathcal{F} \, \mathcal{E}_D^P \, \mathcal{G}$ and $V_\xi = 0$. We conjecture that this cannot occur. A more general generic conjecture can be stated as follows.

Conjecture 6.1. *For any vector a of appropriate order,*

$$(6.1) \qquad\qquad \xi'a \neq 0 \;\Rightarrow\; V_\xi a \neq 0.$$

At present, we are unable to prove or disprove this conjecture, even if $\mathcal{F} \subset \mathcal{G}$. If it does not hold, the consequences are quite far-reaching. Assume that it is possible that $\mathcal{F} \not\mathcal{E}_D^P \mathcal{G}$, i.e., $\xi \neq 0$, but $V_\xi = 0$. Let $\hat{V}_\xi$ converge π_D-almost surely to V_ξ, and let the rank of $\hat{V}_\xi$ converge π_D-almost surely to the rank of V_ξ, i.e., to zero. A standard procedure to test $H_\mathcal{E} : \mathcal{F} \mathcal{E}_D^P \mathcal{G}$ is to reject $H_\mathcal{E}$ if and only if $T\hat{\xi}'\hat{V}_\xi^+\hat{\xi}$ exceeds some (positive) critical value. Many standard inferential procedures fall into this category. Noting that $T\hat{\xi}'\hat{V}_\xi^+\hat{\xi}$ converges π_D-almost surely to zero, the test is seen to be inconsistent for any $\xi \neq 0$. If, however, the conjecture holds, then $\xi \neq 0 \Rightarrow \xi'V_\xi\xi \neq 0 \Rightarrow \xi'V_\xi^+\xi \neq 0$ and $T\hat{\xi}'\hat{V}_\xi^+\hat{\xi}$ converges π_D-almost surely to $+\infty$, implying consistency.

The weighting matrices used to construct quadratic forms in $\hat{\xi}$ are denoted by $\hat{W}_\xi$ or $\hat{V}_\xi^+$. We restrict attention to matrices $\hat{W}_\xi$ and $\hat{V}_\xi$ which satisfy the following two generic conditions. We also need a consistent estimator $\hat{H}_G$ of H_G.

Condition C1. $\hat{W}_\xi$ is symmetric, positive semi-definite and converges π_D-almost surely to W_ξ, a matrix of finite constants.

Condition C2. $\hat{V}_\xi$ is symmetric, positive semi-definite and converges π_D-almost surely to V_ξ.

Condition C3. $\hat{H}_G$ is symmetric, positive definite and converges π_D-almost surely to H_G.

Under Conditions C1–C2, the eigenvalues of $\hat{W}_\xi\hat{V}_\xi$ are real, non-negative, and converge π_D-almost surely to the eigenvalues of $W_\xi V_\xi$. Clearly, a large class of matrices $\hat{W}_\xi$ meets Condition C1 and, as noted before, matrices $\hat{V}_\xi$ and $\hat{H}_G$ satisfying Conditions C2 and C3, respectively, can be constructed from the data. Note also that $\hat{V}_\zeta$ converges π_D-almost surely to V_ζ under Condition C2 and that $\hat{H}_G^{-1}$ converges π_D-almost surely to H_G^{-1} under Condition C3.

We will now define encompassing tests based on the Wald statistics $T\hat{\phi}'\hat{W}_\phi\hat{\phi}$, $T\hat{\phi}^{*\prime}\hat{W}_\phi^*\hat{\phi}^*$, $T\hat{\phi}'\hat{V}_\phi^+\hat{\phi}$ and $T\hat{\phi}^{*\prime}\hat{V}_\phi^{*+}\hat{\phi}^*$, and the score statistics $T\hat{q}'\hat{W}_q\hat{q}$, $T\hat{q}^{*\prime}\hat{W}_q^*\hat{q}^*$, $T\hat{q}'\hat{V}_q^+\hat{q}$ and $T\hat{q}^{*\prime}\hat{V}_q^{*+}\hat{q}^*$, generically denoted by $T\hat{\xi}'\hat{W}_\xi\hat{\xi}$ and $T\hat{\xi}'\hat{V}_\xi^+\hat{\xi}$, and on the the likelihood ratio statistics $-2T\hat{l}$ and $-2T\hat{l}^*$, generically denoted by $-2T\hat{\zeta}$. We denote a test defined by a statistic τ and an associated critical region R_τ (which may be stochastic) by $\langle \tau, R_\tau \rangle$. The test rejects the null if and only if $\tau \in R_\tau$. For any level of significance ϵ

between 0 and 1 and any non-zero vector $\lambda \geq 0$, define the critical value $c_\epsilon(\lambda)$, associated with the distribution $M(\lambda)$, by

$$(6.2) \qquad \Pr[Z \geq c_\epsilon(\lambda)] = \epsilon,$$

where the random variable Z has the distribution $M(\lambda)$. Note that, if $M(\lambda) = \chi_r^2$, then $\lambda = \iota_r$, where ι_r is an $r \times 1$ vector or ones. Let $r = \text{rank}(V_\xi)$. Define the Wald and score tests

$$(6.3) \qquad \langle T\hat{\xi}'\hat{W}_\xi\hat{\xi}, \ [c_\epsilon(\lambda(\hat{W}_\xi\hat{V}_\xi)), +\infty)\rangle,$$

$$(6.4) \qquad \langle T\hat{\xi}'\hat{V}_\xi^+\hat{\xi}, \ [c_\epsilon(\iota_r), +\infty)\rangle,$$

and the likelihood ratio tests

$$(6.5) \qquad \langle -2T\hat{\zeta}, \ [c_\epsilon(\lambda(\hat{H}_G\hat{V}_\zeta)), +\infty)\rangle.$$

We have the following results concerning these tests. We use the definition of $\bar{b}$ in (5.14).

Theorem 6.1. *Given Assumptions A1–A6, B1–B7, N1–N5, Conditions C1– C3, and assuming that $T\bar{b} = o_p(1)$, and that $W_\xi V_\xi \neq 0$ and $V_\zeta \neq 0$, the tests defined by (6.3)–(6.5) are tests of $H_\mathcal{E} : \mathcal{F} \ \mathcal{E}_D^P \ \mathcal{G}$ with asymptotic level ϵ.*

Proof. The result for the test defined by (6.4) follows immediately from Lemma 5.3, part (i). As for (6.3) and (6.5), we need to prove that, if $\mathcal{F} \ \mathcal{E}_D^P \ \mathcal{G}$, then

$$(6.6) \qquad \lim_{T \to \infty} \Pr_{\pi_D}\left[T\hat{\xi}'\hat{W}_\xi\hat{\xi} \geq c_\epsilon(\lambda(\hat{W}_\xi\hat{V}_\xi))\right] = \epsilon$$

and

$$(6.7) \qquad \lim_{T \to \infty} \Pr_{\pi_D}\left[-2T\hat{\zeta} \geq c_\epsilon(\lambda(\hat{H}_G\hat{V}_\zeta))\right] = \epsilon.$$

Clearly, $\hat{W}_\xi\hat{V}_\xi \xrightarrow[\pi_D]{a.s.} W_\xi V_\xi \neq 0$ and $\hat{H}_G\hat{V}_\zeta \xrightarrow[\pi_D]{a.s.} H_G V_\zeta \neq 0$. Hence, given the continuity of $c_\epsilon \circ \lambda$,

$$(6.8) \qquad c_\epsilon(\lambda(\hat{W}_\phi\hat{V}_\phi)) \xrightarrow[\pi_D]{a.s.} c_\epsilon(\lambda(W_\phi V_\phi))$$

92

and

$$(6.9) \qquad c_\epsilon\big(\lambda(\hat{H}_G\hat{V}_\zeta)\big) \xrightarrow[\pi_D]{a.s.} c_\epsilon\big(\lambda(H_G V_\zeta)\big).$$

On the other hand, it follows from Theorems 4.1–4.3 and Lemma 5.2, part (i), that

$$(6.10) \qquad T\hat{\xi}'\hat{W}_\xi\hat{\xi} \xrightarrow[\pi_D]{d} M\big(\lambda(W_\xi V_\xi)\big),$$

$$(6.11) \qquad -2T\hat{\zeta} \xrightarrow[\pi_D]{d} M\big(\lambda(H_G V_\zeta)\big).$$

Then, (6.6)–(6.7) follow from (6.8)–(6.11). $\qquad\square$

Theorem 6.2. *Given Assumptions A1–A6, B1–B7, N1–N5, and Conditions C1–C3, the implicit null associated with (6.3)–(6.5) is characterized by*

$$(6.12) \qquad T\xi'\hat{W}_\xi\xi = O_p(1),$$

$$(6.13) \qquad V_\xi\xi = 0 \quad \text{and} \quad T\bar{b} = O_p(1),$$

and

$$(6.14) \qquad \zeta = 0,$$

respectively.

Proof. The results follow from part (iv) of Lemmas 5.2–5.3 and from Theorem 4.3. $\qquad\square$

Theorem 6.1 asserts that, if $\mathcal{F} \, \mathcal{E}_D^P \, \mathcal{G}$, then the tests (6.3)–(6.5) reject $H_\mathcal{E} : \mathcal{F} \, \mathcal{E}_D^P \, \mathcal{G}$ incorrectly with a probability that goes to ϵ as $T \to \infty$. In other words, (6.3)–(6.5) have correct asymptotic size. Theorem 6.2 shows that (6.3)–(6.5) do not reject $H_\mathcal{E} : \mathcal{F} \, \mathcal{E}_D^P \, \mathcal{G}$ with probability 1 asymptotically if and only if $T\xi'\hat{W}_\xi\xi = O_p(1)$, $V_\xi\xi = 0$ and $T\bar{b} = O_p(1)$, and $\zeta = 0$, respectively. Note that (6.3) can be a consistent test even if $W_\xi\xi = 0$. As yet, since we do not know whether Conjecture 6.1 holds or not, we are unable to give a more basic characterization of the implicit null associated with (6.3)–(6.4).

We note that many standard nested and non-nested tests fall within (6.3)–(6.5), Cox's [1961, 1962] and Vuong's [1989] non-nested likelihood ratio tests being notable exceptions. However, since most procedures derive the asymptotic covariance matrices in a non-robust framework, they generally differ from the correct ones, thereby implying incorrect asymptotic size in general. Exceptions are, for example, Kent [1982] and White [1982] in the context of nested models, and Smith [1993], White [1994] and Gouriéroux–Monfort [1995] in the context of nested and non-nested models.

Finally, many asymptotically equivalent tests under $H_{\mathcal{E}}$ are easily deduced from Theorem 4.4, *inter alia* between likelihood ratio tests and other types of tests.

7. Nested models and implicit form formulation

In practice, one frequently deals with nested models, i.e., where $\mathcal{F} \subset \mathcal{G}$. This case is also of special interest from a theoretical point of view. First, the identity $F_\alpha = G_{\beta_\alpha}$ allows many of the results of the previous section to be simplified. In this respect, it is important to note that $\hat{\phi} = \hat{\phi}^*$, $\hat{q} = \hat{q}^*$ and $\hat{l} = \hat{l}^*$, since $\beta_\alpha = \beta_\alpha^*$ for all $\alpha \in \Omega_{\mathcal{F}}$—see also Gouriéroux–Monfort–Trognon [1983]. Hence, we only have to consider $\hat{\phi}$, $\hat{q}$ and $\hat{l}$. Secondly, under additional regularity conditions, the encompassing hypothesis can be conveniently written in implicit form or, as we will show, in terms of almost sure limits of some statistics which bear a direct relation to the implicit form. These statistics are shown to be asymptotically normally distributed under general conditions.

7.1. Nested models

Assume $\mathcal{F} \subset \mathcal{G}$, and define

$$(7.1) \qquad S(\alpha) = \frac{\partial}{\partial \alpha'} \operatorname{vec} B'(\alpha),$$

$$(7.2) \qquad S = S(\alpha_D),$$

where vec denotes the column expansion of the matrix following it. $S(\alpha)$ is the Hessian matrix of β_α. Furthermore, let I_m be the $m \times m$ identity matrix. The next lemma establishes some links between $\mathcal{F}$ and $\mathcal{G}$. It is useful in deriving some further consequences from Theorems 4.1–4.3.

94

Lemma 7.1. *Let $\mathcal{F} \subset \mathcal{G}$. Given Assumptions A1–A6, B1–B7 and N1–N5,*

$$(7.3) \qquad\qquad L_F^t(\alpha) = L_G^t(\beta_\alpha),$$

$$(7.4) \qquad\qquad 0 = B' q,$$

$$(7.5) \qquad\qquad J_F = B' \tilde{J}_G B,$$

$$(7.6) \qquad\qquad \tilde{J}_{FG} = B' \tilde{J}_G,$$

$$(7.7) \qquad\qquad H_F = B' \tilde{H}_G B - (q' \otimes I_m) S.$$

If furthermore $\mathcal{F} \; \mathcal{E}_D^P \; \mathcal{G}$, then

$$(7.8) \qquad\qquad J_F = B' J_G B,$$

$$(7.9) \qquad\qquad J_{FG} = B' J_G,$$

$$(7.10) \qquad\qquad H_F = B' H_G B.$$

Proof. For all $\alpha \in \Omega_\mathcal{F}$, $F_\alpha = G_{\beta_\alpha}$, wherefrom (7.3) follows. Differentiating (7.3) with respect to α yields

$$(7.11) \qquad\qquad Q_F^t(\alpha) = B'(\alpha) Q_G^t(\beta_\alpha).$$

Evaluating both sides of (7.11) at $\alpha = \alpha_D$ and taking expectations with respect to π_D gives (7.4). A similar operation on the outer vector product of both sides of (7.11) and $Q_F^t(\alpha)$ (resp. $Q_G^t(\beta)$) yields (7.5) (resp. (7.6)). Differentiating (7.11) with respect to α' gives

$$(7.12) \qquad H_F^t(\alpha) = B'(\alpha) H_G^t(\beta_\alpha) B(\alpha) + \left(Q_G^{t\,'}(\beta_\alpha) \otimes I_m \right) S(\alpha).$$

Evaluating (7.12) at $\alpha = \alpha_D$ and taking expectations with respect to π_D gives (7.7). Finally, the equalities (7.8)–(7.10) follow from (7.5)–(7.7) using the properties which led to Corollary 4.1. $\qquad \square$

If $\mathcal{F} \subset \mathcal{G}$, then V_q and V_l can be rewritten as

$$(7.13) \qquad V_q = (I - \tilde{H}_G B H_F^{-1} B') \tilde{J}_G (I - B H_F^{-1} B' \tilde{H}_G),$$

$$(7.14) \qquad\qquad V_l = \omega_G,$$

using (7.4)–(7.6). It appears that V_ϕ cannot be rewritten in a form which is similar to (7.13). Corollary 4.1 takes the following form in case $\mathcal{F} \subset \mathcal{G}$.

Corollary 7.1. *Let $\mathcal{F} \subset \mathcal{G}$, and let Assumptions A1–A6, B1–B7 and N1–N5 hold.*

(i) If $\mathcal{F} \; \mathcal{E}_D^P \; \mathcal{G}$, then

(7.15)
$$\sqrt{T}\hat{\phi} \xrightarrow[\pi_D]{d} N\big(0, \big(H_G^{-1} - B(B'H_GB)^{-1}B'\big)J_G\big(H_G^{-1} - B(B'H_GB)^{-1}B'\big)\big),$$

(7.16)
$$\sqrt{T}\hat{q} \xrightarrow[\pi_D]{d} N\big(0, \big(I - H_GB(B'H_GB)^{-1}B'\big)J_G\big(I - H_GB(B'H_GB)^{-1}B'\big)\big),$$

(7.17)
$$-2T\hat{l} \xrightarrow[\pi_D]{d} M\big(\lambda\big(J_GH_G^{-1} - J_GB(B'H_GB)^{-1}B'\big)\big).$$

(ii) If $D \in \mathcal{F}$, then

(7.18)
$$\sqrt{T}\hat{\phi} \xrightarrow[\pi_D]{d} N\big(0, J_G^{-1} - B(B'J_GB)^{-1}B'\big),$$

(7.19)
$$\sqrt{T}\hat{q} \xrightarrow[\pi_D]{d} N\big(0, J_G - J_GB(B'J_GB)^{-1}B'J_G\big),$$

(7.20)
$$-2T\hat{l} \xrightarrow[\pi_D]{d} \chi^2_{n-m}.$$

Proof. Using Lemma 7.1, (7.15) and (7.16) follow immediately from Corollary 4.1, part (i). Using Theorem 4.3, part (i), and noting that

(7.21)
$$\begin{aligned}\lambda(H_GV_\phi) &= \lambda\big(\big(I - H_GB(B'H_GB)^{-1}B'\big)^2 J_GH_G^{-1}\big)\\ &= \lambda\big(J_GH_G^{-1} - J_GB(B'H_GB)^{-1}B'\big),\end{aligned}$$

we obtain (7.17). Given that $H_G = J_G$ if $D \in \mathcal{F} \subset \mathcal{G}$, (7.18) and (7.19) follow from (7.15) and (7.16). Furthermore, $J_GH_G^{-1} - J_GB(B'H_GB)^{-1}B' = I - J_GB(B'J_GB)^{-1}B'$. This matrix is idempotent with rank equal to $n - m$ and has therefore $n-m$ non-zero eigenvalues, which are all equal to 1. Hence, (7.20) follows from (7.17). $\square$

Notice that Corollary 7.1, part (ii), covers both part (ii) and part (iii) of Corollary 4.1 in case $\mathcal{F} \subset \mathcal{G}$.

7.2. Implicit form formulation

If $\mathcal{F} \subset \mathcal{G}$, then $\mathcal{F} \, \mathcal{E}_D^P \, \mathcal{G}$ if and only if $G_{\beta_D} \in \mathcal{F}$. Hence, $H_\mathcal{E} : \mathcal{F} \, \mathcal{E}_D^P \, \mathcal{G}$ is seen to be a restriction on β_D. Under additional regularity conditions on $\mathcal{F}$ and/or $\mathcal{G}$, $H_\mathcal{E}$ can be equivalently written in implicit form as

$$(7.22) \qquad\qquad H'_\mathcal{E} : h(\beta_D) = 0,$$

for some (non-unique) $r \times 1$ vector-valued function h. Notice that $H_\mathcal{E}$ and $H'_\mathcal{E}$ are equivalent if and only if β_D exists and

$$(7.23) \qquad\qquad h(\beta) = 0 \iff \beta \in \Omega_{\mathcal{G}_\mathcal{F}}.$$

In words, the reflecting set $\Omega_{\mathcal{G}_\mathcal{F}}$ is the parameter space in terms of β associated with $\mathcal{G}_\mathcal{F} = \mathcal{F}$, and the equivalence in (7.23) states that $h(\beta) = 0$ if and only if β belongs to this restricted parameter space. Hence, we proceed under the following assumption.

Assumption I1. *(a)* $\mathcal{F} \subset \mathcal{G}$. *(b)* β_D *exists and is unique. (c)* $\mathcal{F} \, \mathcal{E}_D^P \, \mathcal{G}$ *if and only if* $h(\beta_D) = 0$.

We also make the following assumptions.

Assumption I2. $h(\beta)$ *is twice continuously differentiable on* $\Omega_\mathcal{G}$.

Assumptions I1–I2 ensure the existence of the following matrices:

$$(7.24) \qquad\qquad K = \left. \frac{\partial h(\beta)}{\partial \beta'} \right|_{\beta = \beta_D}$$

and

$$(7.25) \qquad\qquad \tilde{K} = \left. \frac{\partial h(\beta)}{\partial \beta'} \right|_{\beta = \beta_{\alpha_D}}.$$

Assumption I3. *The Jacobian matrix of* $h(\beta)$ *has full row rank* $r = n - m$ *in some open set containing the line segment joining* β_D *and* β_{α_D}.

Note that $h(\beta_\alpha) = 0$ for all $\alpha \in \Omega_\mathcal{F}$. Differentiating this equation with respect to α and evaluating the result at $\alpha = \alpha_D$ yields the orthogonality condition

$$(7.26) \qquad\qquad \tilde{K} B = 0.$$

Under the assumptions made, the matrices $\tilde{K}$ and B are $(n-m) \times m$ and $m \times n$, respectively, and have rank $n-m$ and n, respectively. Hence, the columns of B span the null space of $\tilde{K}$ and the rows of $\tilde{K}$ span the null space of B'. From (7.26), it follows that

$$(7.27) \qquad B(B'H_G B)^{-1} B' + H_G^{-1} \tilde{K}'(\tilde{K} H_G^{-1} \tilde{K}')^{-1} \tilde{K} H_G^{-1} = H_G^{-1},$$

as we will show. This identity will prove useful later on. Define the non-singular matrix

$$(7.28) \qquad F = \begin{pmatrix} \tilde{K} \\ B'H_G \end{pmatrix}.$$

Pre- and postmultiplying both sides of (7.27) by F and F', respectively, yields $F H_G^{-1} F' = F H_G^{-1} F'$. This proves (7.27). Note also that the proof allows H_G to be replaced in (7.27) by any non-singular matrix of the same order.

First, we note that, given the implicit form formulation of the encompassing hypothesis, it is possible to reformulate Corollary 7.1 by means of the matrix K instead of B. Using (7.27) and the fact that $\tilde{K} = K$ if $\mathcal{F} \; \mathcal{E}_D^P \; \mathcal{G}$, we obtain the following.

Corollary 7.2. *Let Assumptions A1–A6, B1–B7, N1–N5 and I1–I3 hold.*
(i) If $\mathcal{F} \; \mathcal{E}_D^P \; \mathcal{G}$, then

$$(7.29)$$
$$\sqrt{T}\hat{\phi} \xrightarrow[\pi_D]{d} N\left(0, H_G^{-1} K'(K H_G^{-1} K')^{-1} K H_G^{-1} J_G H_G^{-1} K'(K H_G^{-1} K')^{-1} K H_G^{-1}\right),$$

$$(7.30) \quad \sqrt{T}\hat{q} \xrightarrow[\pi_D]{d} N\left(0, K'(K H_G^{-1} K')^{-1} K H_G^{-1} J_G H_G^{-1} K'(K H_G^{-1} K')^{-1} K\right),$$

$$(7.31) \qquad -2T\hat{l} \xrightarrow[\pi_D]{d} M\left(\lambda\left((K H_G^{-1} K')^{-1} K H_G^{-1} J_G H_G^{-1} K'\right)\right).$$

(ii) If $D \in \mathcal{F}$, then

$$(7.32) \qquad \sqrt{T}\hat{\phi} \xrightarrow[\pi_D]{d} N\left(0, J_G^{-1} K'(K J_G^{-1} K')^{-1} K J_G^{-1}\right),$$

$$(7.33) \qquad \sqrt{T}\hat{q} \xrightarrow[\pi_D]{d} N\left(0, K'(K J_G^{-1} K')^{-1} K\right),$$

(7.34)
$$-2T\hat{\imath} \xrightarrow[\pi_D]{d} \chi^2_{n-m}.$$

More importantly, it is possible to reformulate $H_{\mathcal{E}}$ in terms of π_D-almost sure limits of some statistics of the data which bear a direct relation to the implicit form formulation of the encompassing hypothesis. As an alternative to ϕ and $\hat{\phi}$, we will consider

$$(7.35) \qquad \eta = h(\beta_D),$$

$$(7.36) \qquad \hat{\eta} = h(\hat{\beta}),$$

and, as an alternative to q and $\hat{q}$, the Lagrange multipliers associated with the constrained maximization of $L_G(\beta)$, which we will discuss now. Given $\mathcal{F} \subset \mathcal{G}$, $\beta_{\hat{\alpha}}$ is seen to be the maximizer of $L_G(\beta)$ over $\Omega_{\mathcal{G}_{\mathcal{F}}}$. This constrained maximum problem may be solved by considering the Lagrangian function

$$(7.37) \qquad \mathcal{L}(\beta, \mu) = L_G(\beta) - \mu' h(\beta),$$

where μ is an $r \times 1$ vector of Lagrange multipliers, and solving $\beta_{\hat{\alpha}}$ and $\mu_{\hat{\alpha}}$ from the conditions

$$(7.38) \qquad \hat{q} - \left.\frac{\partial h'(\beta)}{\partial \beta}\right|_{\beta=\beta_{\hat{\alpha}}} \mu_{\hat{\alpha}} = 0,$$

$$(7.39) \qquad h(\beta_{\hat{\alpha}}) = 0.$$

It is easily seen that $\beta_{\hat{\alpha}}$ and $\mu_{\hat{\alpha}}$ converge π_D-almost surely to β_{α_D} and μ_{α_D}, respectively, which are the solutions of the limit conditions corresponding to (7.38)–(7.39), viz.

$$(7.40) \qquad q - \tilde{K}' \mu_{\alpha_D} = 0,$$

$$(7.41) \qquad h(\beta_{\alpha_D}) = 0.$$

Defining the Lagrange multipliers

$$(7.42) \qquad \mu = \mu_{\alpha_D},$$

and

$$(7.43) \qquad \hat{\mu} = \mu_{\hat{\alpha}},$$

we have the following lemma. It supplements Lemmas 3.1 and 4.4.

Lemma 7.2. *Given Assumptions A1–A4 and I1–I2,*

$$\mathcal{F} \, \mathcal{E}_D^P \, \mathcal{G} \iff \eta = 0 \iff \hat{\eta} \xrightarrow[\pi_D]{a.s.} 0. \tag{7.44}$$

Furthermore, given Assumptions A1–A7, B1–B6 and I1–I3,

$$\mathcal{F} \, \mathcal{E}_D^P \, \mathcal{G} \iff \mu = 0 \iff \hat{\mu} \xrightarrow[\pi_D]{a.s.} 0. \tag{7.45}$$

Proof. The result in (7.44) is immediate, while (7.45) follows from (7.40) and Lemma 3.1. □

The following theorems characterize the limit distributions of $\hat{\eta}$ and $\hat{\mu}$ under general conditions.

Theorem 7.1. *Given Assumptions A1–A6, B1–B7, N1–N5 and I1–I3,*

$$\sqrt{T}(\hat{\eta} - \eta) \xrightarrow[\pi_D]{d} N(0, V_\eta), \tag{7.46}$$

where

$$V_\eta = K H_G^{-1} J_G H_G^{-1} K'. \tag{7.47}$$

Proof. The result follows from the Taylor expansion

$$\sqrt{T}\hat{\eta} = \sqrt{T}\eta + \sqrt{T}K(\hat{\beta} - \beta_D) + o_p(1), \tag{7.48}$$

and from Lemma 4.2. □

Define

$$M = \left[\frac{\partial}{\partial \beta'} \left(\text{vec} \, \frac{\partial h'(\beta)}{\partial \beta} \right) \right]_{\beta=\beta_{\alpha_D}}, \tag{7.49}$$

i.e., M is the Hessian matrix of $h(\beta)$ evaluated at $\beta = \beta_{\alpha_D}$.

Theorem 7.2. *Given Assumptions A1–A6, B1–B7, N1–N5 and I1–I3,*

$$\sqrt{T}(\hat{\mu} - \mu) \xrightarrow[\pi_D]{d} N(0, V_\mu), \tag{7.50}$$

where

$$V_\mu = (\tilde{K}\tilde{K}')^{-1}\tilde{K}\left[I - (\tilde{H}_G + (\mu' \otimes I_m)M)BH_F^{-1}B'\right]\tilde{J}_G \tag{7.51}$$
$$\left[I - BH_F^{-1}B'(\tilde{H}_G + M'(\mu \otimes I_m))\right]\tilde{K}(\tilde{K}\tilde{K}')^{-1}.$$

Proof. From (7.38) and (7.40), it follows that

$$\sqrt{T}(\hat{q} - q) = \sqrt{T}\big(\hat{\tilde{K}}'(\hat{\mu} - \mu) + (\hat{\tilde{K}} - \tilde{K})'\mu\big)$$

$$(7.52) \qquad = \sqrt{T}\big(\tilde{K}'(\hat{\mu} - \mu) + (\hat{\tilde{K}} - \tilde{K})'\mu\big) + o_p(1),$$

where $\hat{\tilde{K}} = [\partial h(\beta)/\partial \beta']_{\beta = \beta_{\hat{a}}}$. Hence,

$$(7.53) \qquad \sqrt{T}(\hat{\mu} - \mu) = \sqrt{T}(\tilde{K}\tilde{K}')^{-1}\tilde{K}\big(\hat{q} - q - (\hat{\tilde{K}} - \tilde{K})'\mu\big) + o_p(1).$$

From the expansion

$$(7.54) \qquad \sqrt{T}\,\text{vec}\,\hat{\tilde{K}}' = \sqrt{T}\,\text{vec}\,\tilde{K}' + \sqrt{T}M(\beta_{\hat{a}} - \beta_{\alpha_D}) + o_p(1)$$

and the fact that $(\hat{\tilde{K}} - \tilde{K})'\mu = (\mu' \otimes I_m)\,\text{vec}(\hat{\tilde{K}} - \tilde{K})'$, we obtain

$$(7.55) \qquad \sqrt{T}(\hat{\tilde{K}} - \tilde{K})'\mu = \sqrt{T}(\mu' \otimes I_m)MBH_F^{-1}Q_F(\alpha_D) + o_p(1),$$

using (4.31). Substituting (4.53) and (7.55) into (7.53) yields

$$(7.56) \qquad \sqrt{T}(\hat{\mu} - \mu) = \sqrt{T}(\tilde{K}\tilde{K}')^{-1}\tilde{K}\Big(Q_G(\beta_{\alpha_D}) - q$$

$$- \big(\tilde{H}_G + (\mu' \otimes I_m)M\big)BH_F^{-1}Q_F(\alpha_D)\Big) + o_p(1).$$

Noting that, from (7.4) and (7.11),

$$(7.57) \qquad Q_F(\alpha_D) = B'\Big(Q_G(\beta_{\alpha_D}) - q\Big),$$

the desired result follows from (7.56). $\qquad\qquad\qquad\qquad\qquad\square$

For any given asymptotic level, consistent Wald and Lagrange multiplier tests of the encompassing hypothesis are easily obtained from Theorems 7.1 and 7.2, respectively. The comments given in the more general context of Section 4 apply here as well.

We obtain the following corollary to Theorems 7.1 and 7.2 under the additional assumptions (i) $\mathcal{F} \, \mathcal{E}_D^P \, \mathcal{G}$ and (ii) $D \in \mathcal{F}$. It follows from Lemma 7.1 and (7.27), and by noting that (i) implies $\eta = \mu = 0$ and $\tilde{K} = K$ and (ii) implies (i) and $H_G = J_G$.

Corollary 7.3. *Let Assumptions A1–A6, B1–B7, N1–N5 and I1–I3 hold.*
(i) If $\mathcal{F} \, \mathcal{E}_D^P \, \mathcal{G}$, *then*

$$(7.58) \qquad \sqrt{T}\hat{\eta} \xrightarrow[\pi_D]{d} N(0, KH_G^{-1}J_GH_G^{-1}K'),$$

$$(7.59) \qquad \sqrt{T}\hat{\mu} \xrightarrow[\pi_D]{d} N\big(0, (KH_G^{-1}K')^{-1}KH_G^{-1}J_GH_G^{-1}K'(KH_G^{-1}K')^{-1}\big).$$

(ii) If $D \in \mathcal{F}$, *then*

$$(7.60) \qquad \sqrt{T}\hat{\eta} \xrightarrow[\pi_D]{d} N(0, KJ_G^{-1}K'),$$

$$(7.61) \qquad \sqrt{T}\hat{\mu} \xrightarrow[\pi_D]{d} N\big(0, (KJ_G^{-1}K')^{-1}\big).$$

Note the similarities between Corollary 7.2 and Corollary 7.3. We end this section by establishing some asymptotic equivalences between $\hat{\phi}$, $\hat{q}$, $\hat{\eta}$ and $\hat{\mu}$, given that $\mathcal{F} \, \mathcal{E}_D^P \, \mathcal{G}$.

Theorem 7.3. *Given Assumptions A1–A6, B1–B7, N1–N5 and I1–I3, if* $\mathcal{F} \, \mathcal{E}_D^P \, \mathcal{G}$, *then*

$$(7.62) \qquad \sqrt{T}\big(\hat{\phi} - H_G^{-1}K'(KH_G^{-1}K')^{-1}\hat{\eta}\big) \xrightarrow[\pi_D]{P} 0,$$

$$(7.63) \qquad \sqrt{T}(\hat{\eta} - K\hat{\phi}) \xrightarrow[\pi_D]{P} 0,$$

$$(7.64) \qquad \sqrt{T}(\hat{q} - K'\hat{\mu}) \xrightarrow[\pi_D]{P} 0,$$

$$(7.65) \qquad \sqrt{T}\big(\hat{\mu} - (KK')^{-1}K\hat{q}\big) \xrightarrow[\pi_D]{P} 0.$$

Proof. Assuming $\mathcal{F} \, \mathcal{E}_D^P \, \mathcal{G}$, it follows from (7.57) that

$$(7.66) \qquad Q_F(\alpha_D) = B'Q_G(\beta_D).$$

Hence, using (4.98), (7.10) and (7.27),

$$(7.67) \qquad \begin{aligned} \sqrt{T}\hat{\phi} &= \sqrt{T}\big(H_G^{-1} - BH_F^{-1}B'\big)Q_G(\beta_D) + o_p(1) \\ &= \sqrt{T}H_G^{-1}K'(KH_G^{-1}K')^{-1}KH_G^{-1}Q_G(\beta_D) + o_p(1). \end{aligned}$$

On the other hand, it follows from (7.48) and (4.29) that also

$$(7.68) \quad \sqrt{T} H_G^{-1} K'(K H_G^{-1} K')^{-1} \hat{\eta} =$$
$$\sqrt{T} H_G^{-1} K'(K H_G^{-1} K')^{-1} K H_G^{-1} Q_G(\beta_D) + o_p(1),$$

which proves (7.62). Premultiplying (7.62) by K yields (7.63). From (4.92) and (7.67), it follows that

$$(7.69) \quad \sqrt{T}\hat{q} = \sqrt{T} K'(K H_G^{-1} K')^{-1} K H_G^{-1} Q_G(\beta_D) + o_p(1).$$

On the other hand, from (7.56), (7.10) and (7.27),

$$\sqrt{T} K' \hat{\mu} = \sqrt{T} K'(K K')^{-1} K H_G$$
$$(H_G^{-1} - B(B' H_G B)^{-1} B') Q_G(\beta_D) + o_p(1)$$
$$(7.70) \qquad = \sqrt{T} K'(K H_G^{-1} K')^{-1} K H_G^{-1} Q_G(\beta_D) + o_p(1),$$

which proves (7.64). Premultiplying (7.64) by $(K K')^{-1} K$ yields (7.65). □

Having derived the limit distributions of $\hat{\eta}$ and $\hat{\mu}$ under general conditions, it is a straightforward matter to construct Wald and Lagrange multiplier encompassing tests based on quadratic forms in $\hat{\eta}$ and $\hat{\mu}$, respectively. In fact, by an appropriate change of the regularity assumptions of Theorems 6.1 and 6.2, the results concerning these tests are subsumed within the statements of those theorems.

8. Conclusion

We have analyzed the possibilities of the Wald, the score and the likelihood ratio principles for testing an encompassing hypothesis of general form. The framework we have adopted is more general than usual in the sense that the models may be nested, disjoint, or overlapping, and that the true distribution may belong to both, only one, or none of the models. It has been shown that a Wald vector, a score vector, and a (modified) likelihood ratio arise as basic statistics for testing the encompassing hypothesis. In particular, the encompassing hypothesis is equivalent to each of these statistics converging almost surely to zero. We obtained the limit distributions of these statistics both under the encompassing hypothesis and under fixed deviations from the encompassing hypothesis. The Wald and score vectors are asymptotically normal in general, while the likelihood ratio has an asymptotic weighted sum of chi-squares distribution under the encompassing hypothesis, and an asymptotic normal distribution otherwise. Similarly, appropriate quadratic forms in the Wald and score vectors have asymptotic weighted sum of chi-squares distributions under the encompassing hypothesis, and asymptotic normal distributions otherwise. A large class of consistent encompassing tests, with correct asymptotic size, have been deduced from these results. By imposing additional assumptions, in particular that the models be nested and/or that the true distribution lies within one of the specified models, we have reobtained many results from the theories of robust nested hypotheses testing and non-robust non-nested hypothesis testing.

This chapter is incomplete in many respects. First, the regularity conditions, which were induced by the methods of proof, are sometimes rather stringent. Weaker smoothness and integrability conditions, still sustaining asymptotic normality, remain to be developed. Secondly, the question whether a zero asymptotic variance of the Wald and score vectors and the likelihood ratio is compatible with a deviation from the encompassing hypothesis is left unanswered. Hopefully, this question can be resolved and lead to a more precise characterization of the implicit null of the tests considered. Third, related to the previous question, no power analysis has been carried out, although the limit results under deviations from the encompassing hypothesis and the limit results concerning quadratic forms are a step in this direction.

Chapter 4

Encompassing tests for the linear model

1. Introduction

The study of linear models together with all the extensions and ramifications has a long history in econometrics, and in statistics in general. Many tools for statistical inference have been designed for, or first been applied to, linear models. This is partly for reasons of analytical and computational tractability, although the latter becomes less and less of a problem. Furthermore, new ideas, unless by nature they pertain to nonlinear models, are more easily introduced and absorbed in the context of linear models. Nonnested hypotheses testing and encompassing are no exception to this rule. The normal linear regression model has indeed been the workhorse in this area.

Pesaran [1974] introduced Cox's [1961, 1962] test for testing non-nested linear regression models. Subsequently, several other tests have been proposed, mostly based on artificial regressions, like the J test of Davidson–MacKinnon [1981], the JA test of Fisher–McAleer [1981] and the F test of Deaton [1982] and Dastoor [1983]. See MacKinnon [1992] for a review on artificial regressions and specification tests. The above-mentioned tests, in particular their power properties and the relations among them in general, have been extensively studied. See Davidson–MacKinnon [1982], Godfrey–Pesaran [1983], MacKinnon [1983], McAleer [1983, 1987], Sawyer [1983], Ericsson [1986], McAleer–Pesaran [1986], Pesaran [1987], Dastoor–McAleer [1989], Zabel [1993], and Fan–Li [1995], among others. McAleer [1995] discusses non-nested hypotheses testing in applied econometric work.

In Chapter 3 we have drawn a distinction between robust and non-robust hypothesis testing, and have proposed to reserve the term 'encompassing' for the former category—recall Figure 3.1. As we argued there, most of the existing literature on hypothesis testing, nested and non-nested alike, is to be qualified as non-robust, including most of the work that actually goes under the heading 'encompassing'. Notable exceptions are Smith

[1993], White [1994], Gouriéroux–Monfort [1994, 1995], and, in a model selection context, Vuong [1989]. In line with these authors, we adopt a robust approach towards the problem of testing non-nested regression models. The distinctive feature, compared to the standard Cox, J, JA and F tests, is that we do not assume that the model under test, $\mathcal{F}$, is correctly specified. Instead, it is only assumed that $\mathcal{F}$ encompasses the 'alternative' model $\mathcal{G}$. It turns out that a comprehensive model which nests both $\mathcal{F}$ and $\mathcal{G}$ plays the same role as in standard non-nested tests, but the statistics obtained here are different, and their asymptotic distribution holds under more general conditions. In particular we show that a robust test, in the sense of White [1982], of $\mathcal{F}$ against the comprehensive model is in fact an encompassing test.

The models and the true distribution are introduced in Section 2. In Section 3, pseudo-true values, pseudo-true disturbances and related quantities are defined and calculated. There, also, the encompassing relation is interpreted in several equivalent ways, which all revolve around some decomposition of L_2. The limit distributions of the Wald and score vectors and the modified likelihood ratio are derived in Section 4. Section 5 and part of Section 6 are devoted to the sample analogues of the quantities encountered earlier. Encompassing tests are then derived, and some asymptotic equivalences are noted. Section 6 closes with a discussion of standard non-nested testing procedures from the viewpoint of encompassing. Section 7 concludes this chapter.

2. The framework

The general framework is that of Chapter 3, Section 3, where the models $\mathcal{F}$ and $\mathcal{G}$ are univariate normal linear regression models with constant variance, as in Examples 1.5.3 and 2.6.3. The details are as follows. We consider a random scalar Y and a random vector X taking values y in R and x in R^l, respectively. Let P denote the marginal distribution of X, D the conditional distribution of Y, given X, and π_D the joint distribution of Y and X. The models are defined as $\mathcal{F} = \{F_\alpha | \alpha \in \Omega_{\mathcal{F}}\}$ and $\mathcal{G} = \{G_\beta | \beta \in \Omega_{\mathcal{G}}\}$, with

$$(2.1) \qquad F_\alpha = N(X_F'\alpha_1, \alpha_2), \qquad (\alpha_1, \alpha_2) \in R^m \times R_+ = \Omega_{\mathcal{F}},$$

and

$$(2.2) \qquad G_\beta = N(X_G'\beta_1, \beta_2), \qquad (\beta_1, \beta_2) \in R^n \times R_+ = \Omega_{\mathcal{G}},$$

i.e., by means of the measurable functions X_F and X_G of X, defining the 'regressors'. X_F and X_G take values x_F in R^m and x_G in R^n, respectively. The conditional density functions associated with F_α and G_β are given by

$$(2.3) \qquad f(y|x;\alpha) = \frac{1}{\sqrt{2\pi\alpha_2}} \exp\left[-\frac{(y - x_F'\alpha_1)^2}{2\alpha_2}\right]$$

and

$$(2.4) \qquad g(y|x;\beta) = \frac{1}{\sqrt{2\pi\beta_2}} \exp\left[-\frac{(y - x_G'\beta_1)^2}{2\beta_2}\right],$$

respectively.

By formulating $\mathcal{F}$, the modeller expresses his view that the true conditional distribution of Y, given X, is in some sense or for some purpose well approximated by a normal distribution whose conditional mean is some linear function of X_F (which in itself is a function of X) and whose conditional variance does not depend on X. A similar interpretation pertains to $\mathcal{G}$.

We make the following assumption.

Assumption A1. (a) Y is square integrable with respect to π_D. (b) The components of X_F and X_G are square integrable with respect to P, and $E_P(X_F X_F')$ and $E_P(X_G X_G')$ are non-singular.

Let $\mu = \mu(X)$ and $\sigma^2 = \sigma^2(X)$ denote the true conditional mean and variance, respectively, of Y, given X. That is, $\mu = E_D(Y|X)$ and $\sigma^2 = E_D((Y-\mu)^2|X)$. Under Assumption A1-(a), μ and σ^2 exist P-almost surely, and $E_P(\mu^2)$ and $E_P(\sigma^2)$ exist. Furthermore, under Assumption A1-(b), the dimensions of the linear subspaces of $L_2(R^l, P)$ spanned by the components of X_F and of X_G are m and n, respectively. In other words, $\mathcal{F}$ and $\mathcal{G}$ are linear models of full rank.

There exists a unique minimal linear model which nests both $\mathcal{F}$ and $\mathcal{G}$, that is, a linear model with minimal dimension which has $\mathcal{F}$ and $\mathcal{G}$ as submodels. To show this, let $m + r$ be the dimension of the joint span of X_F and X_G, i.e.,

$$(2.5) \qquad r = \text{rank}\, E_P \begin{pmatrix} X_F X_F' & X_F X_G' \\ X_G X_F' & X_G X_G' \end{pmatrix} - m.$$

Obviously, $n \geq r$. We shall exclude the trivial case $r = 0$ or, equivalently, $\mathcal{G} \subset \mathcal{F}$. Indeed it will be shown that $\mathcal{G} \subset \mathcal{F}$ implies $\mathcal{F} \; \mathcal{E}_D^P \; \mathcal{G}$, so no testing would be required. Let $X_\cup$ be a measurable function of X, taking values $x_\cup$ in R^{m+r}, whose components form an orthonormal basis of the joint span of X_F and X_G. Then

(2.6)
$$E_P(X_\cup X'_\cup) = I,$$

and

(2.7)
$$X_F = E_P(X_F X'_\cup) X_\cup,$$

(2.8)
$$X_G = E_P(X_G X'_\cup) X_\cup,$$

P-almost surely. $X_\cup$ is obtained by stacking the regressors X_F and X_G, removing linear dependencies among them, followed by an orthogonalization and normalization. The resulting $X_\cup$ is a set of regressors defining the minimal linear nesting model of $\mathcal{F}$ and $\mathcal{G}$. Though $X_\cup$ is non-unique (it may be premultiplied by an orthogonal matrix), the induced minimal linear nesting model is unique and, by construction, of full rank. An explicit expression for $X_\cup$ may be found as follows. Consider the decomposition

(2.9)
$$E_P \begin{pmatrix} X_F X'_F & X_F X'_G \\ X_G X'_F & X_G X'_G \end{pmatrix} = S \Lambda S',$$

where S is an $(m+n) \times (m+r)$ semi-orthogonal matrix and Λ an $(m+r) \times (m+r)$ positive definite diagonal matrix. (The diagonal elements of Λ are the non-zero eigenvalues of the matrix on the LHS of (2.9), and the columns of S the corresponding normalized eigenvectors.) It is easily seen, then, that (2.6)–(2.8) are satisfied if we put

(2.10)
$$X_\cup = \Lambda^{-1/2} S' \begin{pmatrix} X_F \\ X_G \end{pmatrix},$$

for example.

We need to introduce some notation for various subspaces of $L_2(R^l, P)$. We shall denote the span of, say X_F, by $\mathcal{S}(X_F)$, its orthogonal complement in $L_2(R^l, P)$ by $\mathcal{S}^\perp(X_F)$, and its orthogonal complement in $\mathcal{S}(X_\cup)$ by $\mathcal{S}^\perp_{X_\cup}(X_F)$. Observe that $\mathcal{S}(X_F)$ and $\mathcal{S}^\perp_{X_\cup}(X_F)$ have dimension m and r, respectively. As we shall clarify later on, encompassing has to do with a property of the decomposition of μ into its orthogonal projections on $\mathcal{S}^\perp(X_\mathsf{U})$, $\mathcal{S}(X_F)$ and $\mathcal{S}^\perp_{X_\mathsf{U}}(X_F)$. It is therefore helpful to think in terms of these subspaces rather than in terms of (non-invariant) parameters. Define $F = E_P(X_F X'_\mathsf{U})$, $G = E_P(X_G X'_\mathsf{U})$, $P_F = F'(FF')^{-1}F$, $P_G = G'(GG')^{-1}G$, $M_F = I - P_F$ and $M_G = I - P_G$. Observe that F and G have full row rank m and n, respectively, that P_F, P_G, M_F and M_G are idempotent, and that $M_F P_F = 0$ and $M_G P_G = 0$. Let $\mu_X = E_P(X_\mathsf{U}\mu)$, i.e., μ_X is the vector of coordinates in the basis X_U of the projection of μ onto $\mathcal{S}(X_\mathsf{U})$. The above definitions will allow recurrent expressions like $E_P(\mu X'_F)[E_P(X_F X'_F)]^{-1} E_P(X_F \mu)$ to be simplified to $\mu'_X P_F \mu_X$.

3. Pseudo-true values and other limits

In this section, we derive the non-random quantities which are directly related to the encompassing hypothesis. First, we calculate the pseudo-true values α_D, β_D and $\beta_{\alpha D}$ for the models considered and discuss the encompassing hypothesis, thereby slightly generalizing the conclusion reached in Example 2.6.3. Secondly, we compute the difference ϕ, the score q, and the modified log-likelihood ratio l, which have been defined in Chapter 3, Section 3.

Much of the ensuing discussion is based on the log-density functions $\log f$ and $\log g$ and their first and second derivatives. These functions are given by

$$(3.1) \qquad \log f(y|x;\alpha) = -\frac{1}{2}\log(2\pi) - \frac{1}{2}\log\alpha_2 - \frac{(y - x'_F\alpha_1)^2}{2\alpha_2},$$

$$(3.2) \qquad \frac{\partial \log f(y|x;\alpha)}{\partial \alpha} = \begin{pmatrix} \dfrac{x_F(y - x'_F\alpha_1)}{\alpha_2} \\ \dfrac{(y - x'_F\alpha_1)^2 - \alpha_2}{2\alpha_2^2} \end{pmatrix},$$

and

$$(3.3) \qquad \frac{\partial^2 \log f(y|x;\alpha)}{\partial\alpha\partial\alpha'} = -\begin{pmatrix} \dfrac{x_F x'_F}{\alpha_2} & \dfrac{x_F(y - x'_F\alpha_1)}{\alpha_2^2} \\ \dfrac{(y - x'_F\alpha_1)x'_F}{\alpha_2^2} & \dfrac{2(y - x'_F\alpha_1)^2 - \alpha_2}{2\alpha_2^3} \end{pmatrix},$$

with obvious analogues for $\log g$ and its derivatives.

3.1. Pseudo-true values and the encompassing hypothesis

We start by taking expectations of (3.1)–(3.2) with respect to π_D, giving

$$(3.4) \qquad E_P E_D \log f(Y|X;\alpha) = -\frac{1}{2}\log(2\pi) - \frac{1}{2}\log\alpha_2$$
$$-\frac{E_P\sigma^2 + E_P(\mu - X_F'\alpha_1)^2}{2\alpha_2}$$

and

$$(3.5) \qquad E_P E_D \frac{\partial \log f(Y|X;\alpha)}{\partial\alpha} = \left(\begin{array}{c} \dfrac{E_P(X_F\mu) - E_P(X_F X_F')\alpha_1}{\alpha_2} \\[2ex] \dfrac{E_P\sigma^2 + E_P(\mu - X_F'\alpha_1)^2 - \alpha_2}{2\alpha_2^2} \end{array}\right).$$

Equating the RHS of (3.5) to zero and solving for α yields the pseudo-true value

$$(3.6) \quad \alpha_D = \left(\begin{array}{c} [E_P(X_F X_F')]^{-1} E_P(X_F\mu) \\[1ex] E_P\sigma^2 + E_P\mu^2 - E_P(\mu X_F')[E_P(X_F X_F')]^{-1} E_P(X_F\mu) \end{array}\right)$$

$$= \left(\begin{array}{c} (FF')^{-1}F\mu_X \\[1ex] E_P\sigma^2 + E_P\mu^2 - \mu_X' P_F\mu_X \end{array}\right).$$

Similarly,

$$(3.7) \quad \beta_D = \left(\begin{array}{c} [E_P(X_G X_G')]^{-1} E_P(X_G\mu) \\[1ex] E_P\sigma^2 + E_P\mu^2 - E_P(\mu X_G')[E_P(X_G X_G')]^{-1} E_P(X_G\mu) \end{array}\right)$$

$$= \left(\begin{array}{c} (GG')^{-1}G\mu_X \\[1ex] E_P\sigma^2 + E_P\mu^2 - \mu_X' P_G\mu_X \end{array}\right).$$

Substituting F_α for D yields

$$(3.8) \quad \beta_\alpha = \left(\begin{array}{c} [E_P(X_G X_G')]^{-1} E_P(X_G m_1) \\[1ex] \alpha_2 + E_P(m_1^2) - E_P(m_1 X_G')[E_P(X_G X_G')]^{-1} E_P(X_G m_1) \end{array}\right).$$

where $m_1 = X_F'\alpha_1$. Recall (1.5.21). Hence, putting $\alpha = \alpha_D$,

$$(3.9) \quad \beta_{\alpha_D} = \left(\begin{array}{c} [E_P(X_G X_G')]^{-1} E_P(X_G X_F')[E_P(X_F X_F')]^{-1} E_P(X_F\mu) \\[1ex] E_P\sigma^2 + E_P\mu^2 - E_P(\mu X_F')[E_P(X_F X_F')]^{-1} E_P(X_F X_G') \\ [E_P(X_G X_G')]^{-1} E_P(X_G X_F') \\ [E_P(X_F X_F')]^{-1} E_P(X_F\mu) \end{array}\right)$$

$$= \left(\begin{array}{c} (GG')^{-1}GP_F\mu_X \\[1ex] E_P\sigma^2 + E_P\mu^2 - \mu_X' P_F P_G P_F\mu_X \end{array}\right).$$

Assumption A1 ensures the existence and uniqueness of α_D, β_D and β_{α_D}. Note that (3.7) and (3.9) generalize (2.6.7) and (2.6.8) to the case where D is arbitrary instead of normal with constant variance. Remark also that, in obvious notation, β_{2D} is a function of β_{1D} and that $\beta_{2\alpha_D}$ is the same function of $\beta_{1\alpha_D}$:

$$(3.10) \qquad \beta_{2D} = E_P\sigma^2 + E_P\mu^2 - \beta'_{1D}E_P(X_GX'_G)\beta_{1D},$$

$$(3.11) \qquad \beta_{2\alpha_D} = E_P\sigma^2 + E_P\mu^2 - \beta'_{1\alpha_D}E_P(X_GX'_G)\beta_{1\alpha_D}.$$

Hence

$$(3.12) \qquad \beta_{1D} = \beta_{1\alpha_D} \Rightarrow \beta_{2D} = \beta_{2\alpha_D},$$

the reverse implication not being true in general. Interestingly, the sign of $\beta_{2D} - \beta_{2\alpha_D} = \mu'_X(P_FP_GP_F - P_G)\mu_X$ is not determined a priori, since $P_FP_GP_F - P_G$ has no (semi-)definite character in general.

Comparing (3.7) with (3.9) yields exactly the same conclusion as in Example 2.6.3, i.e.,

$$(3.13) \quad \mathcal{F} \, \mathcal{E}_D^P \, \mathcal{G} \iff E_P(X_G\mu) = E_P(X_GX'_F)\left[E_P(X_FX'_F)\right]^{-1} E_P(X_F\mu)$$
$$\iff GM_F\mu_X = 0.$$

Equivalently, $\mathcal{F} \, \mathcal{E}_D^P \, \mathcal{G}$ iff X_G and μ are orthogonal, given X_F.

There is also a direct interpretation of encompassing in terms of the projection of μ onto the subspaces $\mathcal{S}(X_F)$ and $\mathcal{S}(X_\cup)$. From (3.13),

$$(3.14) \quad \mathcal{F} \, \mathcal{E}_D^P \, \mathcal{G} \iff E_P(X_\cup\mu) = E_P(X_\cup X'_F)\left[E_P(X_FX'_F)\right]^{-1} E_P(X_F\mu).$$

Premultiplying by $X'_\cup$ and using (2.7), it follows that

$$(3.15) \quad \mathcal{F} \, \mathcal{E}_D^P \, \mathcal{G} \iff$$
$$X'_\cup E_P(X_\cup\mu) = X'_F\left[E_P(X_FX'_F)\right]^{-1} E_P(X_F\mu) \quad P-\text{a.s.}$$

Hence, $\mathcal{F} \, \mathcal{E}_D^P \, \mathcal{G}$ iff the projection of μ onto $\mathcal{S}(X_\cup)$ is equal to that onto $\mathcal{S}(X_F)$. Furthermore, since $L_2(R^l, P)$ is the direct sum of $\mathcal{S}^\perp(X_\cup)$, $\mathcal{S}(X_F)$

and $S_{X_U}^\perp(X_F)$, μ can be decomposed into mutually orthogonal components as

(3.16)
$$\mu = \mu_U^\perp + \mu_F + \mu_F^\perp,$$

where $\mu_U^\perp$, μ_F and $\mu_F^\perp$ are the orthogonal projections of μ onto $S^\perp(X_U)$, $S(X_F)$ and $S_{X_U}^\perp(X_F)$, respectively. It follows that $\mathcal{F}\ \mathcal{E}_D^P\ \mathcal{G}$ iff $\mu_F^\perp = 0$. An equivalent formulation is that $\mathcal{F}\ \mathcal{E}_D^P\ \mathcal{G}$ iff $\mu_U \in S(X_F)$ (P-almost surely) in the decomposition $\mu = \mu_U^\perp + \mu_U$. Note furthermore that $\mu_F^\perp$ is a point in an r-dimensional function space for which we can always find an orthonormal basis such that only one coordinate of $\mu_F^\perp$ would be non-zero if $\mathcal{F}\ \not\mathcal{E}_D^P\ \mathcal{G}$. This illustrates the fact that incomplete encompassing, as defined in Chapter 2, Section 3, is not invariant with respect to the parametrization chosen and that, although encompassing boils down to an orthogonality condition between μ and some r-dimensional space, we cannot unambiguously speak of 'the number of violations of the encompassing hypothesis'. The only thing that matters is the orthogonality or non-orthogonality of μ and $S_{X_U}^\perp(X_F)$.

The decomposition in (3.16) allows one to draw some further conclusions. First, observe that $\mathcal{F}\ \mathcal{E}_D^P\ \mathcal{G}$ whenever $\mathcal{G} \subset \mathcal{F}$. Secondly, considering any set of regressors $X_{G'}$ associated with a model $\mathcal{G}'$ such that the joint span of X_F and X_G equals (resp. contains) that of X_F and $X_{G'}$, we have $\mathcal{F}\ \mathcal{E}_D^P\ \mathcal{G} \iff \mathcal{F}\ \mathcal{E}_D^P\ \mathcal{G}'$ (resp. $\mathcal{F}\ \mathcal{E}_D^P\ \mathcal{G} \Rightarrow \mathcal{F}\ \mathcal{E}_D^P\ \mathcal{G}'$). In particular, let $\mathcal{G}'$ be the model induced by a set of r regressors $X_{G'}$ such that $S(X_{G'}) = S_{X_U}^\perp(X_F)$. Then $\mathcal{F}\ \mathcal{E}_D^P\ \mathcal{G} \iff \mathcal{F}\ \mathcal{E}_D^P\ \mathcal{G}'$, showing that the regressors in X_G which are linear functions of those in X_F may be eliminated from X_G without changing the encompassing hypothesis under study. Alternatively, let $\mathcal{G}'$ be the minimal linear nesting model induced by X_F and X_G. Then $\mathcal{F}\ \mathcal{E}_D^P\ \mathcal{G} \iff \mathcal{F}\ \mathcal{E}_D^P\ \mathcal{G}'$. In view of this equivalence, $\mathcal{G}$ might be replaced by the minimal nesting model $\mathcal{G}'$ in the statistical procedures which follow. This would reduce a problem of testing non-nested models to one of testing nested models without altering the implicit null hypothesis, thereby considerably simplifying the algebra. In some sense this finding also renders much of the existing literature on non-nested testing in normal linear models redundant.

3.2. Pseudo-true disturbances

It is convenient to introduce the 'disturbances'

$$(3.17) \qquad U_F = Y - X_F'\alpha_{1D} = Y - X_U'P_F\mu_X,$$

$$(3.18) \qquad U_G = Y - X_G'\beta_{1D} = Y - X_U'P_G\mu_X,$$

$$(3.19) \qquad U_{\tilde{G}} = Y - X_G'\beta_{1\alpha_D} = Y - X_U'P_G P_F\mu_X,$$

$$(3.20) \qquad U_{\tilde{G}} = X_F'\alpha_{1D} - X_G'\beta_{1\alpha_D} = U_{\tilde{G}} - U_F = X_U'M_G P_F\mu_X.$$

The first three of these may rightly be called pseudo-true disturbances, whereas $U_{\tilde{G}}$ is equal to $E_{\alpha_D}U_{\tilde{G}}$. The disturbances allow the first order conditions defining α_D, β_D and β_{α_D} to be written as

$$(3.21) \qquad E_P E_D(X_F U_F) = 0, \qquad\qquad E_P E_D(U_F^2 - \alpha_{2D}) = 0,$$

$$(3.22) \qquad E_P E_D(X_G U_G) = 0, \qquad\qquad E_P E_D(U_G^2 - \beta_{2D}) = 0,$$

$$(3.23) \qquad E_P E_{\alpha_D}(X_G U_{\tilde{G}}) = 0, \qquad\qquad E_P E_{\alpha_D}(U_{\tilde{G}}^2 - \beta_{2\alpha_D}) = 0.$$

The last pair of conditions is equivalent to

$$(3.24) \qquad E_P(X_G U_{\tilde{G}}) = 0, \qquad\qquad E_P(U_{\tilde{G}}^2 + \alpha_{2D} - \beta_{2\alpha_D}) = 0.$$

Then, noting that

$$(3.25) \qquad \beta_{1D} - \beta_{1\alpha_D} = [E_P(X_G X_G')]^{-1} E_P E_D(X_G U_F),$$

it follows that $\mathcal{F}\ \mathcal{E}_D^P\ \mathcal{G}$ iff

$$(3.26) \qquad E_P E_D(X_G U_F) = 0.$$

In other words, $\mathcal{F}\ \mathcal{E}_D^P\ \mathcal{G}$ iff $E_D U_F$ (which is the conditional mean misspecification of $\mathcal{F}$) is orthogonal to X_G. Note that, if $D \in \mathcal{F}$, then $E_D U_F = 0$ and hence $\mathcal{F}\ \mathcal{E}_D^P\ \mathcal{G}$.

3.3. Other equivalents of the encompassing hypothesis

Consider now ϕ, q and l, which, we recall, are defined as

$$(3.27) \qquad\qquad \phi = \beta_D - \beta_{\alpha_D},$$

$$(3.28) \qquad\qquad q = E_P E_D \left[\frac{\partial \log g(Y|X; \beta)}{\partial \beta} \right]_{\beta = \beta_{\alpha_D}},$$

$$(3.29) \qquad l = E_P E_D \left[\log g(Y|X; \beta_{\alpha_D}) - \log g(Y|X; \beta_D) \right].$$

Observe that ϕ, q and l exist and are unique under Assumption A1. There are various ways to proceed. For example, ϕ can be computed straightforwardly from (3.7) and (3.9), and q and l from the analogues to (3.5) and (3.4), respectively. This gives, after some algebra,

$$(3.30) \qquad\qquad \phi = \begin{pmatrix} (GG')^{-1} G M_F \mu_X \\ \mu_X'(P_F P_G P_F - P_G)\mu_X \end{pmatrix},$$

$$(3.31) \qquad q = \begin{pmatrix} \dfrac{G M_F \mu_X}{E_P \sigma^2 + E_P \mu^2 - \mu_X' P_F P_G P_F \mu_X} \\[2mm] -\dfrac{\mu_X' P_F P_G M_F \mu_X}{\left[E_P \sigma^2 + E_P \mu^2 - \mu_X' P_F P_G P_F \mu_X \right]^2} \end{pmatrix},$$

$$
\begin{aligned}
(3.32) \qquad l = &-\frac{1}{2} \log \frac{E_P \sigma^2 + E_P \mu^2 - \mu_X' P_F P_G P_F \mu_X}{E_P \sigma^2 + E_P \mu^2 - \mu_X' P_G \mu_X} \\
&+ \frac{\mu_X' P_F P_G M_F \mu_X}{E_P \sigma^2 + E_P \mu^2 - \mu_X' P_F P_G P_F \mu_X}.
\end{aligned}
$$

We shall also express ϕ, q and l as functions of β_{1D}, $\beta_{1\alpha_D}$, β_{2D} and $\beta_{2\alpha_D}$, in a way which more closely resembles their empirical counterparts $\hat{\phi}$, $\hat{q}$ and $\hat{l}$, which will be discussed in Section 5. From (3.7), it follows that

$$(3.33) \qquad E_P(X_G \mu) - E_P(X_G X_G') \beta_{1\alpha_D} = E_P(X_G X_G')(\beta_{1D} - \beta_{1\alpha_D}),$$

which, combined with (3.11), gives in turn

$$
\begin{aligned}
(3.34) \qquad E_P \sigma^2 + E_P &\left(\mu - X_G' \beta_{1\alpha_D} \right)^2 - \beta_{2\alpha_D} \\
&= -2\beta_{1\alpha_D}' E_P(X_G X_G')(\beta_{1D} - \beta_{1\alpha_D}).
\end{aligned}
$$

Using (3.33)–(3.34) in the analogues to (3.4)–(3.5), it now follows easily that

$$(3.35) \qquad \phi = \begin{pmatrix} \beta_{1D} - \beta_{1\alpha D} \\ \beta_{2D} - \beta_{2\alpha D} \end{pmatrix},$$

$$(3.36) \qquad q = \begin{pmatrix} \dfrac{E_P(X_G X_G')(\beta_{1D} - \beta_{1\alpha D})}{\beta_{2\alpha D}} \\[2ex] -\dfrac{\beta_{1\alpha D}' E_P(X_G X_G')(\beta_{1D} - \beta_{1\alpha D})}{\beta_{2\alpha D}^2} \end{pmatrix},$$

$$(3.37) \qquad l = -\frac{1}{2} \log \frac{\beta_{2\alpha D}}{\beta_{2D}} + \frac{\beta_{1\alpha D}' E_P(X_G X_G')(\beta_{1D} - \beta_{1\alpha D})}{\beta_{2\alpha D}}.$$

Given Assumption A1, it is easy to see that $\mathcal{F} \, \mathcal{E}_D^P \, \mathcal{G} \iff \phi = 0 \iff q = 0 \iff l = 0 \iff \phi_1 = 0 \iff q_1 = 0$, in obvious notation.

Alternative expressions for q and l can be deduced directly from the analogues to (3.1)–(3.2), giving

$$(3.38) \qquad q = \begin{pmatrix} \dfrac{E_P E_D(X_G U_{\tilde{G}})}{\beta_{2\alpha D}} \\[2ex] \dfrac{E_P E_D(U_{\tilde{G}}^2) - \beta_{2\alpha D}}{2\beta_{2\alpha D}^2} \end{pmatrix},$$

$$(3.39) \qquad l = -\frac{1}{2} \log \frac{\beta_{2\alpha D}}{\beta_{2D}} - \frac{E_P E_D(U_{\tilde{G}}^2) - \beta_{2\alpha D}}{2\beta_{2\alpha D}}.$$

Finally, noting that, from (3.20) and (3.24),

$$(3.40) \qquad E_P E_D(X_G U_{\tilde{G}}) = E_P E_D(X_G U_F),$$

$$(3.41) \qquad E_P E_D(U_{\tilde{G}}^2) - \beta_{2\alpha D} = 2 E_P E_D(U_{\tilde{G}} U_F),$$

q and l may also be written as

$$(3.42) \qquad q = \begin{pmatrix} \dfrac{E_P E_D(X_G U_F)}{\beta_{2\alpha D}} \\[2ex] \dfrac{E_P E_D(U_{\tilde{G}} U_F)}{\beta_{2\alpha D}^2} \end{pmatrix},$$

$$(3.43) \qquad l = -\frac{1}{2} \log \frac{\beta_{2\alpha D}}{\beta_{2D}} - \frac{E_P E_D(U_{\tilde{G}} U_F)}{\beta_{2\alpha D}}.$$

4. Limit distributions

Assume that we are given a sample of observations (y_t, x_t), $t = 1, \ldots, T$, which are generated independently by π_D. In the next section, we shall compute the pseudo-ML estimators $\hat{\alpha}$, $\hat{\beta}$ and $\beta^*_{\hat{\alpha}}$, the Wald vector $\hat{\phi}^*$, the score vector $\hat{q}^*$ and the modified log-likelihood ratio $\hat{l}^*$, which have been defined in Chapter 3. Since in this chapter we are only concerned with conditional pseudo-true values (which is the only feasible way to proceed when $\mathcal{F} \not\subset \mathcal{G}$), we shall condense the notation and write β_α, $\beta_{\hat{\alpha}}$, $\hat{\phi}$, $\hat{q}$ and $\hat{l}$ instead of the more burdensome β^*_α, $\beta^*_{\hat{\alpha}}$, $\hat{\phi}^*$, $\hat{q}^*$ and $\hat{l}^*$. This section, applying the results of Chapter 3, gives some limit distribution theory for $\hat{\phi}$, $\hat{q}$ and $\hat{l}$ in the context of linear models. Starting from the asymptotic expansions given there, we derive the first order limit behaviour of $\hat{\phi}$, $\hat{q}$ and $\hat{l}$, which is then specialized to the case where $\mathcal{F} \, \mathcal{E}^P_D \, \mathcal{G}$. This is the first step in a classical approach towards testing $H_\mathcal{E}$. The next step will be carried out in Section 6. It consists of finding consistent estimators of the limit covariance matrices of $\hat{\phi}$ and $\hat{q}$ that attain in finite samples their almost sure limit rank and, finally, of constructing Moore-Penrose inverted quadratic forms in $\sqrt{T}\hat{\phi}$ and $\sqrt{T}\hat{q}$. Under $H_\mathcal{E}$ then, the resulting Wald and score statistics have a χ^2_r distribution, whereas the modified log-likelihood ratio $-2T\hat{l}$ has a weighted sum of chi-squares distribution.

4.1. Preliminaries

Let x_{Ft}, x_{Gt} and x_{Ut}, $t = 1, \ldots, T$, be the realizations of X_F, X_G and X_U, respectively, corresponding to x_t. Then, the components of the log-likelihood function, the score function and the derivative of the score function associated with $\mathcal{F}$ are given by

$$(4.1) \qquad L^t_F(\alpha) = -\frac{1}{2}\log(2\pi) - \frac{1}{2}\log \alpha_2 - \frac{(y_t - x'_{Ft}\alpha_1)^2}{2\alpha_2},$$

$$(4.2) \qquad Q^t_F(\alpha) = \begin{pmatrix} \dfrac{x_{Ft}(y_t - x'_{Ft}\alpha_1)}{\alpha_2} \\[2mm] \dfrac{(y_t - x'_{Ft}\alpha_1)^2 - \alpha_2}{2\alpha_2^2} \end{pmatrix},$$

and

$$(4.3) \qquad H^t_F(\alpha) = -\begin{pmatrix} \dfrac{x_{Ft}x'_{Ft}}{\alpha_2} & \dfrac{x_{Ft}(y_t - x'_{Ft}\alpha_1)}{\alpha_2^2} \\[3mm] \dfrac{(y_t - x'_{Ft}\alpha_1)x'_{Ft}}{\alpha_2^2} & \dfrac{2(y_t - x'_{Ft}\alpha_1)^2 - \alpha_2}{2\alpha_2^3} \end{pmatrix},$$

respectively, with obvious analogues for $L_G^t(\beta)$ and its derivatives.

We shall now compute the matrices H_F, H_G, $\tilde{H}_G$, $\bar{H}_G$ and B, which appear in the asymptotic expansions (3.4.45) and (3.4.54). Recall the definitions (3.4.5), (3.4.6), (3.4.10), (3.4.14) and (3.4.21). Observing that, from (3.17)–(3.20) and (3.40)–(3.41),

$$(4.4) \qquad E_P E_D(X_G U_{\tilde{G}}) = G M_F \mu_X,$$

$$(4.5) \qquad E_P E_D(2U_{\tilde{G}}^2 - \beta_{\alpha_D}) = \beta_{\alpha_D} - 4\mu_X' P_F P_G M_F \mu_X,$$

$$(4.6) \qquad E_P E_D(U_{\tilde{G}} X_F') = \mu_X' P_F M_G F',$$

we find, using the first order conditions (3.21)–(3.23),

$$(4.7) \qquad H_F = E_P E_D \begin{pmatrix} \dfrac{X_F X_F'}{\alpha_{2D}} & \dfrac{X_F U_F}{\alpha_{2D}^2} \\[2ex] \dfrac{U_F X_F'}{\alpha_{2D}^2} & \dfrac{2U_F^2 - \alpha_{2D}}{2\alpha_{2D}^3} \end{pmatrix} = \begin{pmatrix} \dfrac{FF'}{\alpha_{2D}} & 0 \\[2ex] 0 & \dfrac{1}{2\alpha_{2D}^2} \end{pmatrix},$$

$$(4.8) \qquad H_G = E_P E_D \begin{pmatrix} \dfrac{X_G X_G'}{\beta_{2D}} & \dfrac{X_G U_G}{\beta_{2D}^2} \\[2ex] \dfrac{U_G X_G'}{\beta_{2D}^2} & \dfrac{2U_G^2 - \beta_{2D}}{2\beta_{2D}^3} \end{pmatrix} = \begin{pmatrix} \dfrac{GG'}{\beta_{2D}} & 0 \\[2ex] 0 & \dfrac{1}{2\beta_{2D}^2} \end{pmatrix},$$

$$(4.9) \qquad \tilde{H}_G = E_P E_D \begin{pmatrix} \dfrac{X_G X_G'}{\beta_{2\alpha_D}} & \dfrac{X_G U_{\tilde{G}}}{\beta_{2\alpha_D}^2} \\[2ex] \dfrac{U_{\tilde{G}} X_G'}{\beta_{2\alpha_D}^2} & \dfrac{2U_{\tilde{G}}^2 - \beta_{2\alpha_D}}{2\beta_{2\alpha_D}^3} \end{pmatrix}$$

$$= \begin{pmatrix} \dfrac{GG'}{\beta_{2\alpha_D}} & \dfrac{G M_F \mu_X}{\beta_{2\alpha_D}^2} \\[2ex] \dfrac{\mu_X' M_F G'}{\beta_{2\alpha_D}^2} & \dfrac{\beta_{2\alpha_D} - 4\mu_X' P_F P_G M_F \mu_X}{2\beta_{2\alpha_D}^3} \end{pmatrix},$$

$$(4.10) \qquad \bar{H}_G = E_P E_{\alpha_D} \begin{pmatrix} \dfrac{X_G X_G'}{\beta_{2\alpha_D}} & \dfrac{X_G U_{\tilde{G}}}{\beta_{2\alpha_D}^2} \\[2ex] \dfrac{U_{\tilde{G}} X_G'}{\beta_{2\alpha_D}^2} & \dfrac{2U_{\tilde{G}}^2 - \beta_{2\alpha_D}}{2\beta_{2\alpha_D}^3} \end{pmatrix} = \begin{pmatrix} \dfrac{GG'}{\beta_{2\alpha_D}} & 0 \\[2ex] 0 & \dfrac{1}{2\beta_{2\alpha_D}^2} \end{pmatrix}.$$

Recall from (3.4.22) that $B = \bar{H}_G^{-1} \bar{J}_{GF}$. Thus we need to compute $\bar{J}_{GF}$. Noting that

$$(4.11) \qquad E_{\alpha_D}(U_{\tilde{G}} U_F) = \alpha_{2D},$$

$$(4.12) \qquad E_{\alpha_D}\left[U_{\tilde{G}}(U_F^2 - \alpha_{2D})\right] = 0,$$

$$(4.13) \qquad E_{\alpha_D}\left[(U_{\tilde{G}}^2 - \beta_{2\alpha_D})U_F\right] = 2\alpha_{2D}U_{\tilde{G}},$$

$$(4.14) \qquad E_{\alpha_D}\left[(U_{\tilde{G}}^2 - \beta_{2\alpha_D})(U_F^2 - \alpha_{2D})\right] = 2\alpha_{2D}^2,$$

we have

$$(4.15) \quad \bar{J}_{GF} = E_P E_{\alpha_D} \begin{pmatrix} \dfrac{X_G U_{\tilde{G}} U_F X_F'}{\alpha_{2D}\beta_{2\alpha_D}} & \dfrac{X_G U_{\tilde{G}}(U_F^2 - \alpha_{2D})}{2\alpha_{2D}^2\beta_{2\alpha_D}} \\ \dfrac{(U_{\tilde{G}}^2 - \beta_{2\alpha_D})U_F X_F'}{2\alpha_{2D}\beta_{2\alpha_D}^2} & \dfrac{(U_{\tilde{G}}^2 - \beta_{2\alpha_D})(U_F^2 - \alpha_{2D})}{4\alpha_{2D}^2\beta_{2\alpha_D}^2} \end{pmatrix}$$

$$= E_P \begin{pmatrix} \dfrac{X_G X_F'}{\beta_{2\alpha_D}} & 0 \\ \dfrac{U_{\tilde{G}} X_F'}{\beta_{2\alpha_D}^2} & \dfrac{1}{2\beta_{2\alpha_D}^2} \end{pmatrix} = \begin{pmatrix} \dfrac{GF'}{\beta_{2\alpha_D}} & 0 \\ \dfrac{\mu_X' P_F M_G F'}{\beta_{2\alpha_D}^2} & \dfrac{1}{2\beta_{2\alpha_D}^2} \end{pmatrix}.$$

Consequently,

$$(4.16) \quad B = \begin{pmatrix} [E_P(X_G X_G')]^{-1} E_P(X_G X_F') & 0 \\ 2E_P(U_{\tilde{G}} X_F') & 1 \end{pmatrix} = \begin{pmatrix} (GG')^{-1}GF' & 0 \\ 2\mu_X' P_F M_G F' & 1 \end{pmatrix}.$$

Observe that all of H_F, H_G, $\tilde{H}_G$, $\bar{H}_G$ and B exist under Assumption A1.

For the purpose of computing the vectors $Q_F^t(\alpha_D)$, $Q_G^t(\beta_D)$, $Q_G^t(\beta_{\alpha_D})$ and $E_{\alpha_D}\left[Q_G^t(\beta_{\alpha_D})\right]$, whose sums appear in (3.4.45) and (3.4.54), let u_{Ft}, u_{Gt}, $u_{\tilde{G}t}$ and $u_{\bar{G}t}$, $t = 1, \ldots, T$, be the realizations of U_F, U_G, $U_{\tilde{G}}$ and $U_{\bar{G}}$, respectively, corresponding to (y_t, x_t). That is, $u_{Ft} = y_t - x_{Ft}' \alpha_{1D}$, etc. Then, it follows easily that

$$(4.17) \qquad Q_F^t(\alpha_D) = \begin{pmatrix} \dfrac{x_{Ft} u_{Ft}}{\alpha_{2D}} \\ \dfrac{u_{Ft}^2 - \alpha_{2D}}{2\alpha_{2D}^2} \end{pmatrix} = \begin{pmatrix} \dfrac{F x_{\cup t} u_{Ft}}{\alpha_{2D}} \\ \dfrac{u_{Ft}^2 - \alpha_{2D}}{2\alpha_{2D}^2} \end{pmatrix},$$

118

$$(4.18) \qquad Q_G^t(\beta_D) = \begin{pmatrix} \dfrac{x_{Gt} u_{Gt}}{\beta_{2D}} \\ \dfrac{u_{Gt}^2 - \beta_{2D}}{2\beta_{2D}^2} \end{pmatrix} = \begin{pmatrix} \dfrac{G x_{\cup t} u_{Gt}}{\beta_{2D}} \\ \dfrac{u_{Gt}^2 - \beta_{2D}}{2\beta_{2D}^2} \end{pmatrix},$$

$$(4.19) \qquad Q_G^t(\beta_{\alpha_D}) = \begin{pmatrix} \dfrac{x_{Gt} u_{\tilde{G}t}}{\beta_{2\alpha_D}} \\ \dfrac{u_{\tilde{G}t}^2 - \beta_{2\alpha_D}}{2\beta_{2\alpha_D}^2} \end{pmatrix} = \begin{pmatrix} \dfrac{G x_{\cup t} u_{\tilde{G}t}}{\beta_{2\alpha_D}} \\ \dfrac{u_{\tilde{G}t}^2 - \beta_{2\alpha_D}}{2\beta_{2\alpha_D}^2} \end{pmatrix},$$

$$(4.20) \qquad E_{\alpha_D}\left[Q_G^t(\beta_{\alpha_D})\right] = \begin{pmatrix} \dfrac{x_{Gt} u_{\tilde{G}t}}{\beta_{2\alpha_D}} \\ \dfrac{u_{\tilde{G}t}^2 + \alpha_{2D} - \beta_{2\alpha_D}}{2\beta_{2\alpha_D}^2} \end{pmatrix}$$
$$= \begin{pmatrix} \dfrac{G x_{\cup t}(u_{\tilde{G}t} - u_{Ft})}{\beta_{2\alpha_D}} \\ \dfrac{(u_{\tilde{G}t} - u_{Ft})^2 + \alpha_{2D} - \beta_{2\alpha_D}}{2\beta_{2\alpha_D}^2} \end{pmatrix}.$$

Finally, observe that, under $H_{\mathcal{E}}$,

$$(4.21) \qquad \tilde{H}_G = \bar{H}_G = H_G,$$

$$(4.22) \qquad B = \begin{pmatrix} (GG')^{-1}GF' & 0 \\ 2\mu_X' M_G F' & 1 \end{pmatrix},$$

$$(4.23) \qquad Q_G^t(\beta_{\alpha_D}) = Q_G^t(\beta_D),$$

$$(4.24) \qquad E_{\alpha_D}\left[Q_G^t(\beta_{\alpha_D})\right] = \begin{pmatrix} \dfrac{G x_{\cup t}(u_{Gt} - u_{Ft})}{\beta_{2D}} \\ \dfrac{(u_{Gt} - u_{Ft})^2 + \alpha_{2D} - \beta_{2D}}{2\beta_{2D}^2} \end{pmatrix}.$$

119

4.2. Limit distribution of the Wald vector

Let W to be the vector of components of X_G projected onto $\mathcal{S}^\perp_{X_\cup}(X_F)$, i.e.

$$(4.25) \qquad W = X_G - E_P(X_G X_F')\left[E_P(X_F X_F')\right]^{-1} X_F,$$

and let w_t be the realization of W corresponding to x_t. We will show that, under $H_{\mathcal{E}}$, the leading term in the expansion of $\hat{\phi}$ is a linear transformation of $T^{-1}\sum_{t=1}^T w_t u_{Ft}$. Hence, the random variable WU_F is crucial in determining the limit behaviour of $\sqrt{T}\hat{\phi}$. As shown in the next subsections, the same applies to $\sqrt{T}\hat{q}$ and $-2T\hat{l}$.

Recalling (3.4.45), we find, using (3.20) and the results from the previous subsection,

$$
\begin{aligned}
\sqrt{T}(\hat{\phi} - \phi) &= -\sqrt{T}BH_F^{-1}Q_F(\alpha_D) + \sqrt{T}H_G^{-1}Q_G(\beta_D) \\
&\qquad - \sqrt{T}\bar{H}_G^{-1}E_{\alpha_D}\left[Q_G(\beta_{\alpha_D})\right] + o_p(1) \\
(4.26) \qquad &= \frac{1}{\sqrt{T}}\begin{pmatrix} (GG')^{-1}G \\ -2\mu_X' P_G \end{pmatrix} M_F \sum_{t=1}^T x_{\cup t} u_{Ft} + \frac{R_\phi}{\sqrt{T}} + o_p(1),
\end{aligned}
$$

where

$$(4.27) \quad R_\phi = \sum_{t=1}^T \begin{pmatrix} (GG')^{-1}G x_{\cup t}(u_{Gt} - u_{\tilde{G}t}) \\ 2\mu_X' M_F P_G M_F x_{\cup t} u_{Ft} + u_{Gt}^2 - u_{\tilde{G}t}^2 + \beta_{2\alpha_D} - \beta_{2D} \end{pmatrix}.$$

Observing that $R_\phi = 0$ under $H_{\mathcal{E}}$, we shall study the first term on the RHS of (4.26) in more detail. Recalling expressions like $GG' = E_P(X_G X_G')$, $GX_\cup = X_G$ (P-almost surely) and $\mu_X' G' = E_P(\mu X_G')$, it follows easily that

$$(4.28) \qquad \begin{pmatrix} (GG')^{-1}G \\ -2\mu_X' P_G \end{pmatrix} M_F x_{\cup t} u_{Ft} = A_\phi w_t u_{Ft},$$

where

$$(4.29) \qquad A_\phi = \begin{pmatrix} [E_P(X_G X_G')]^{-1} \\ -2E_P(\mu X_G')[E_P(X_G X_G')]^{-1} \end{pmatrix} = \begin{pmatrix} (GG')^{-1} \\ -2\beta_{1D}' \end{pmatrix}.$$

Remark that, since

$$(4.30) \quad E_P E_D(WU_F) = E_P(X_G\mu) - E_P(X_G X_F')\left[E_P(X_F X_F')\right]^{-1} E_P(X_F\mu),$$

we have $E_P E_D(WU_F) = 0$ iff $\mathcal{F} \; \mathcal{E}_D^P \; \mathcal{G}$. The following assumption is made to ensure the asymptotic normality of $\sqrt{T}\hat{\phi}$.

Assumption A2. (a) Y is fourth order integrable with respect to π_D. (b) The components of X_F and X_G are fourth order integrable with respect to P.

Given Assumption 2, the second moment matrix $\Omega = E_P E_D(W U_F^2 W')$ exists by Hölder's inequality and is given by

$$(4.31) \qquad \Omega = E_P \left[\left(\sigma^2 + (\mu - X_F' \alpha_{1D})^2 \right) W W' \right].$$

Note that, by construction, the n components of W span $\mathcal{S}_{X_U}^{\perp}(X_F)$, whose dimension is r. Hence, the non-degeneracy of D is sufficient to ensure that the rank of Ω is equal to r.

Assumption A3. $\sigma^2 > 0$ P-almost surely.

From the preceding and from the multivariate central limit theorem it follows that, under $H_{\mathcal{E}}$,

$$(4.32) \qquad \sqrt{T}\hat{\phi} \xrightarrow[\pi_D]{d} N(0, V_\phi),$$

where

$$(4.33) \qquad V_\phi = A_\phi \Omega A_\phi'.$$

Clearly, and not surprisingly, $\text{rank}(V_\phi) = r$, since $\mathcal{F}\,\mathcal{E}_D^P\,\mathcal{G}$ is equivalent to μ being orthogonal to the r-dimensional space $\mathcal{S}_{X_U}^{\perp}(X_F)$.

4.3. Limit distribution of the score vector

After some algebra, it follows from expansion (3.4.54) that

$$\sqrt{T}(\hat{q} - q) = -\sqrt{T}\tilde{H}_G B H_F^{-1} Q_F(\alpha_D) + \sqrt{T}\left(Q_G(\beta_{\alpha_D}) - q \right)$$
$$- \sqrt{T}\tilde{H}_G \bar{H}_G^{-1} E_{\alpha_D} \left[Q_G(\beta_{\alpha_D}) \right] + o_p(1)$$

$$(4.34) \qquad = \frac{1}{\sqrt{T}} \begin{pmatrix} \beta_{2\alpha_D}^{-1} G \\ -\beta_{2\alpha_D}^{-2} \mu_X' P_G \end{pmatrix} M_F \sum_{t=1}^{T} x_{Ut} u_{Ft} + \frac{R_q}{\sqrt{T}} + o_p(1),$$

where

$$(4.35)$$
$$R_q = \sum_{t=1}^{T} \begin{pmatrix} -\beta_{2\alpha_D}^{-2} G M_F \mu_X \left(2\mu_X' P_F P_G M_F x_{Ut} u_{Ft} + u_{\tilde{G}t}^2 \right) \\ \beta_{2\alpha_D}^{-2} \mu_X' M_F P_G \left(2 M_F x_{Ut} u_{Ft} - x_{Ut} u_{\tilde{G}t} - P_F \mu_X \right) \\ + 2\beta_{2\alpha_D}^{-3} \mu_X' M_F P_G P_F \mu_X \left(2\mu_X' P_F P_G M_F x_{Ut} u_{Ft} + u_{\tilde{G}t}^2 \right) \end{pmatrix}.$$

121

Under $H_\mathcal{E}$, $R_q = 0$. Defining

$$(4.36) \qquad A_q = \begin{pmatrix} \beta_{2\alpha_D}^{-1} I_n \\ -\beta_{2\alpha_D}^{-2} E_P(\mu X_G') \left[E_P(X_G X_G') \right]^{-1} \end{pmatrix} = \begin{pmatrix} \beta_{2\alpha_D}^{-1} I_n \\ -\beta_{2\alpha_D}^{-2} \beta_{1D}' \end{pmatrix},$$

it follows that

$$(4.37) \qquad \begin{pmatrix} \beta_{2\alpha_D}^{-1} G \\ -\beta_{2\alpha_D}^{-2} \mu_X' P_G \end{pmatrix} M_{F \mathcal{X}_{Ut} u_{Ft}} = A_q w_t u_{Ft}.$$

Consequently, under $H_\mathcal{E}$,

$$(4.38) \qquad \sqrt{T} \hat{q} \xrightarrow[\pi_D]{d} N(0, V_q),$$

where

$$(4.39) \qquad V_q = A_q \Omega A_q'.$$

Again, $\text{rank}(V_q) = r$. Observe furthermore that, if we consider (4.29), (4.33), (4.36) and (4.39) as mere definitions, V_q and V_ϕ are related by

$$(4.40) \qquad V_q = \Delta H_G V_\phi H_G \Delta,$$

where

$$(4.41) \qquad \Delta = \begin{pmatrix} b I_n & 0 \\ 0 & b^2 \end{pmatrix},$$

with the scalar $b = \beta_{2D}/\beta_{2\alpha_D}$ equal to one under $H_\mathcal{E}$, thereby confirming Corollary 3.4.1, part (i). Note that, when $\mathcal{F} \not\mathcal{L}_D^P \mathcal{G}$, either of $b < 1$, $b = 1$ and $b > 1$ are possible.

4.4. Limit distribution of the modified likelihood ratio

From the discussion in Chapter 3, we know that the term in $\sqrt{T}$ of the expansion of $\sqrt{T}(\hat{l} - l)$ is identically zero under $H_\mathcal{E}$—see (3.4.73). If desired, it could be derived in a way similar to R_ϕ and R_q. We shall, however, directly proceed under $H_\mathcal{E}$. Then, from Theorem 3.4.3,

$$(4.42) \qquad -2T\hat{l} \xrightarrow[\pi_D]{d} M \left(\lambda (H_G A_\phi \Omega A_\phi') \right).$$

Since $H_G A_\phi \Omega A'_\phi$ and $A'_\phi H_G A_\phi \Omega$ have the same r non-zero eigenvalues, it follows that

(4.43) $$-2T\hat{l} \xrightarrow[\pi_D]{d} M\left(\lambda(V_l)\right),$$

where

(4.44) $$V_l = A_l \Omega$$

and

(4.45) $$A_l = A'_\phi H_G A_\phi = \frac{(GG')^{-1}}{\beta_{2D}} + 2\frac{\beta_{1D}\beta'_{1D}}{\beta_{2D}^2}.$$

This completes the discussion of the first order asymptotic behaviour of $\hat{\phi}$, $\hat{q}$ and $\hat{l}$ under $H_\mathcal{E}$. Consistent estimators of Ω, A_ϕ, A_q and A_l are given in Section 6.

4.5. Special cases

Here we briefly discuss the main simplifications that occur when (i) $D \in \mathcal{F}$, (ii) $\mathcal{F} \subset \mathcal{G}$, and (iii) $D \in \mathcal{F} \subset \mathcal{G}$.

The study of the limit behaviour of test statistics is usually performed under the assumption of correct specification, i.e., $D \in \mathcal{F}$, which is considerably stronger than $H_\mathcal{E}$, maintained in the previous subsection. The essential simplification brought about by assuming $D \in \mathcal{F}$ concerns Ω, which reduces to

(4.46) $\Omega = \sigma^2 \left[E_P(X_G X'_G) - E_P(X_G X'_F) \left[E_P(X_F X'_F)\right]^{-1} E_P(X_F X'_G) \right]$
$$= \sigma^2 G M_F G',$$

with σ^2 being constant. The limit covariance matrices $A_\phi \Omega A'_\phi$ and $A_q \Omega A'_q$, and $\lambda(A_l \Omega)$ simplify accordingly. Observe that Assumption A2 is no longer required for the existence of Ω. Assumptions A1 and A3 are necessary and sufficient for $\sqrt{T}\hat{\phi}$ and $\sqrt{T}\hat{q}$ to be asymptotically normally distributed, and for $-2T\hat{l}$ to be asymptotically distributed as a weighted sum of chi-squares.

Let $\mathcal{F} \subset \mathcal{G}$. Of the many simplifications that occur in the various expressions for ϕ, q and l, it suffices to note that (3.30)–(3.32) simplify to

(4.47) $$\phi = \begin{pmatrix} (GG')^{-1} G M_F \mu_X \\ \mu'_X (P_F - P_G) \mu_X \end{pmatrix},$$

123

$$(4.48) \qquad q = \begin{pmatrix} \dfrac{GM_F\mu_X}{E_P\sigma^2 + E_P\mu^2 - \mu_X'P_F\mu_X} \\ 0 \end{pmatrix},$$

and

$$(4.49) \qquad l = -\frac{1}{2}\log\frac{E_P\sigma^2 + E_P\mu^2 - \mu_X'P_F\mu_X}{E_P\sigma^2 + E_P\mu^2 - \mu_X'P_G\mu_X},$$

in view of $P_GP_F = P_F$. Observe that now

$$(4.50) \qquad \beta_{1D} = \beta_{1\alpha_D} \iff \beta_{2D} = \beta_{2\alpha_D}$$

and

$$(4.51) \qquad \beta_{2D} - \beta_{2\alpha_D} = -\mu_X'P_GM_FP_G\mu_X \leq 0$$

—compare with (3.12) and the ensuing comments. In short, $\mathcal{F}\,\mathcal{E}_D^P\,\mathcal{G} \iff \phi_1 = 0 \iff \phi_2 = 0 \iff q_1 = 0 \iff l = 0$. Assume that $\mathcal{F}\,\mathcal{E}_D^P\,\mathcal{G}$ in addition to $\mathcal{F} \subset \mathcal{G}$. Then $\beta_{1D}'\Omega = 0$, since $\beta_{1D}'W = 0$. As a result, V_ϕ, V_q and V_l reduce to

$$(4.52) \qquad V_\phi = \begin{pmatrix} (GG')^{-1}\Omega(GG')^{-1} & 0 \\ 0 & 0 \end{pmatrix},$$

$$(4.53) \qquad V_q = \begin{pmatrix} \beta_{2\alpha_D}^{-2}\Omega & 0 \\ 0 & 0 \end{pmatrix},$$

and

$$(4.54) \qquad V_l = \beta_{2D}^{-1}(GG')^{-1}\Omega,$$

respectively.

Finally, assume that $D \in \mathcal{F} \subset \mathcal{G}$. Combining the results from the preceding paragraphs yields

$$(4.55) \qquad V_\phi = \begin{pmatrix} \sigma^2(GG')^{-1}GM_FG'(GG')^{-1} & 0 \\ 0 & 0 \end{pmatrix},$$

$$(4.56) \qquad V_q = \begin{pmatrix} \sigma^{-2}GM_FG' & 0 \\ 0 & 0 \end{pmatrix},$$

and

$$(4.57) \qquad V_l = (GG')^{-1}GM_FG',$$

using the fact that $\beta_{2D} = \beta_{2\alpha_D} = \sigma^2$. Note that, since V_l has $n - m$ non-zero eigenvalues, which are all equal to one, $-2T\hat{l}$ is asymptotically distributed as a χ^2_{n-m} variate, a familiar result.

5. Pseudo-ML estimators and basic statistics

In this section, we compute the pseudo-ML estimators $\hat{\alpha}$, $\hat{\beta}$ and $\beta_{\hat{\alpha}}$, and the statistics $\hat{\phi}$, $\hat{q}$ and $\hat{l}$, all corresponding to the sample of observations (y_t, x_t), $t = 1, \ldots, T$, generated independently by π_D. The development is largely parallel to that of Section 3, with the understanding that here the observations replace the random variables and empirical averages replace expectations. Most of the resulting expressions have a familiar appearance and are therefore presented without much discussion.

5.1. Pseudo-ML estimators

The normalized likelihood and score functions are found by taking averages of their components, given by (4.1)–(4.2). This yields

$$(5.1) \qquad L_F(\alpha) = -\frac{1}{2}\log(2\pi) - \frac{1}{2}\log\alpha_2 - \frac{1}{T}\frac{(\eta - \Xi_F\alpha_1)'(\eta - \Xi_F\alpha_1)}{2\alpha_2}$$

and

$$(5.2) \qquad Q_F(\alpha) = \frac{1}{T}\begin{pmatrix} \dfrac{\Xi_F'(\eta - \Xi_F\alpha_1)}{\alpha_2} \\[2mm] \dfrac{(\eta - \Xi_F\alpha_1)'(\eta - \Xi_F\alpha_1) - T\alpha_2}{2\alpha_2^2} \end{pmatrix},$$

where $\Xi_F = (x_{F1}, \ldots, x_{FT})'$, $\Xi_G = (x_{G1}, \ldots, x_{GT})'$ and $\eta = (y_1, \ldots, y_T)'$. For $T \geq m + r$, we make the following assumption.

Assumption A4. *With P-probability 1, the matrices Ξ_F, Ξ_G and (Ξ_F, Ξ_G) have rank m, n and $m + r$, respectively.*

Solving $Q_F(\alpha) = 0$ with respect to α yields the unique pseudo-ML estimator

$$(5.3) \qquad \hat{\alpha} = \begin{pmatrix} (\Xi_F'\Xi_F)^{-1}\Xi_F'\eta \\[1mm] \frac{1}{T}\eta'(I - \Pi_F)\eta \end{pmatrix},$$

where $\Pi_F = \Xi_F(\Xi_F'\Xi_F)^{-1}\Xi_F'$. Similarly,

$$(5.4) \qquad \hat{\beta} = \begin{pmatrix} (\Xi_G'\Xi_G)^{-1}\Xi_G'\eta \\[1mm] \frac{1}{T}\eta'(I - \Pi_G)\eta \end{pmatrix}.$$

where $\Pi_G = \Xi_G(\Xi_G'\Xi_G)^{-1}\Xi_G'$. Recall from (1.5.27) that

$$(5.5) \qquad \beta_\alpha = \begin{pmatrix} (\Xi_G'\Xi_G)^{-1}\Xi_G'\Xi_F\alpha_1 \\[1mm] \alpha_2 + \frac{1}{T}\alpha_1'\Xi_F'(I - \Pi_G)\Xi_F\alpha_1 \end{pmatrix}.$$

125

Hence, putting $\alpha = \hat{\alpha}$,

(5.6)
$$\beta_{\hat{\alpha}} = \begin{pmatrix} (\Xi'_G \Xi_G)^{-1} \Xi'_G \Pi_F \eta \\ \frac{1}{T} \eta' (I - \Pi_F \Pi_G \Pi_F) \eta \end{pmatrix}.$$

Remark that

(5.7)
$$\hat{\beta}_2 = \frac{1}{T} [\eta'\eta - \hat{\beta}'_1 \Xi'_G \Xi_G \hat{\beta}_1]$$

and

(5.8)
$$\beta_{2\hat{\alpha}} = \frac{1}{T} [\eta'\eta - \beta'_{1\hat{\alpha}} \Xi'_G \Xi_G \beta_{1\hat{\alpha}}].$$

Finally, by analogy to (3.17)–(3.20), define the $T \times 1$ vectors of 'residuals'

(5.9)
$$\hat{u}_F = \eta - \Xi_F \hat{\alpha}_1,$$

(5.10)
$$\hat{u}_G = \eta - \Xi_G \hat{\beta}_1,$$

(5.11)
$$\hat{u}_{\bar{G}} = \eta - \Xi_G \beta_{1\hat{\alpha}},$$

(5.12)
$$\hat{u}_{\tilde{G}} = \Xi_F \hat{\alpha}_1 - \Xi_G \beta_{1\hat{\alpha}} = \hat{u}_{\bar{G}} - \hat{u}_F,$$

and let $\hat{u}_{Ft}$, $\hat{u}_{Gt}$, $\hat{u}_{\tilde{G}t}$ and $\hat{u}_{\bar{G}t}$ denote the t-th element of $\hat{u}_F$, $\hat{u}_G$, $\hat{u}_{\tilde{G}}$ and $\hat{u}_{\bar{G}}$, respectively. Then, the first order conditions defining $\hat{\alpha}$, $\hat{\beta}$ and $\beta_{\hat{\alpha}}$ imply

(5.13)
$$\Xi'_F \hat{u}_F = 0, \qquad \hat{u}'_F \hat{u}_F - T\hat{\alpha}_2 = 0,$$

(5.14)
$$\Xi'_G \hat{u}_G = 0, \qquad \hat{u}'_G \hat{u}_G - T\hat{\beta}_2 = 0,$$

(5.15)
$$\Xi'_G \hat{u}_{\bar{G}} = 0, \qquad \hat{u}'_{\bar{G}} \hat{u}_{\bar{G}} + T\hat{\alpha}_2 - T\beta_{2\hat{\alpha}} = 0.$$

5.2. Wald and score vectors and the modified likelihood ratio

Recall that $\hat{\phi} = \hat{\beta} - \beta_{\hat{\alpha}}$, $\hat{q} = Q_G(\beta_{\hat{\alpha}})$, and $\hat{l} = L_G(\beta_{\hat{\alpha}}) - L_G(\hat{\beta})$. From the analogues to (5.1)–(5.2), we find

$$(5.16) \qquad \hat{\phi} = \begin{pmatrix} (\Xi_G'\Xi_G)^{-1}\,\Xi_G'(I - \Pi_F)\eta \\ \frac{1}{T}\eta'\,(\Pi_F\Pi_G\Pi_F - \Pi_G)\,\eta \end{pmatrix},$$

$$(5.17) \qquad \hat{q} = \begin{pmatrix} \dfrac{\Xi_G'(I - \Pi_F)\eta}{\eta'\,(I - \Pi_F\Pi_G\Pi_F)\,\eta} \\[2ex] -\dfrac{\eta'\Pi_F\Pi_G\,(I - \Pi_F)\,\eta}{\frac{1}{T}\,[\eta'\,(I - \Pi_F\Pi_G\Pi_F)\,\eta]^2} \end{pmatrix},$$

$$(5.18) \qquad \hat{l} = -\frac{1}{2}\log\frac{\eta'\,(I - \Pi_F\Pi_G\Pi_F)\,\eta}{\eta'\,(I - \Pi_G)\,\eta} + \frac{\eta'\Pi_F\Pi_G\,(I - \Pi_F)\,\eta}{\eta'\,(I - \Pi_F\Pi_G\Pi_F)\,\eta}.$$

Compare with (3.30)–(3.32). Furthermore, observing that

$$(5.19) \qquad \Xi_G'\eta - \Xi_G'\Xi_G\beta_{1\hat{\alpha}} = \Xi_G'\Xi_G\big(\hat{\beta}_1 - \beta_{1\hat{\alpha}}\big),$$

$$(5.20) \qquad (\eta - \Xi_G\beta_{1\hat{\alpha}})'(\eta - \Xi_G\beta_{1\hat{\alpha}}) - T\beta_{2\hat{\alpha}} = -2\beta_{1\hat{\alpha}}'\Xi_G'\Xi_G\big(\hat{\beta}_1 - \beta_{1\hat{\alpha}}\big),$$

one may alternatively write

$$(5.21) \qquad \hat{\phi} = \begin{pmatrix} \hat{\beta}_1 - \beta_{1\hat{\alpha}} \\ \hat{\beta}_2 - \beta_{2\hat{\alpha}} \end{pmatrix},$$

$$(5.22) \qquad \hat{q} = \frac{1}{T}\begin{pmatrix} \dfrac{\Xi_G'\Xi_G\big(\hat{\beta}_1 - \beta_{1\hat{\alpha}}\big)}{\beta_{2\hat{\alpha}}} \\[2ex] -\dfrac{\beta_{1\hat{\alpha}}'\Xi_G'\Xi_G\big(\hat{\beta}_1 - \beta_{1\hat{\alpha}}\big)}{\beta_{2\hat{\alpha}}^2} \end{pmatrix},$$

$$(5.23) \qquad \hat{l} = -\frac{1}{2}\log\frac{\beta_{2\hat{\alpha}}}{\hat{\beta}_2} + \frac{1}{T}\frac{\beta_{1\hat{\alpha}}'\Xi_G'\Xi_G\big(\hat{\beta}_1 - \beta_{1\hat{\alpha}}\big)}{\beta_{2\hat{\alpha}}}.$$

Note the similarity with (3.35)–(3.37). This similarity goes further. From the analogues to (5.3)–(5.4), it follows immediately that

$$(5.24) \qquad \hat{q} = \frac{1}{T}\begin{pmatrix} \dfrac{\Xi_G'\hat{u}_{\tilde{G}}}{\beta_{2\hat{\alpha}}} \\[2ex] \dfrac{\hat{u}_{\tilde{G}}'\hat{u}_{\tilde{G}} - T\beta_{2\hat{\alpha}}}{2\beta_{2\hat{\alpha}}^2} \end{pmatrix},$$

127

$$\text{(5.25)} \qquad \hat{l} = -\frac{1}{2}\log\frac{\beta_{2\hat{\alpha}}}{\hat{\beta}_2} + \frac{1}{T}\frac{\hat{u}'_{\tilde{G}}\hat{u}_{\tilde{G}} - T\beta_{2\hat{\alpha}}}{2\beta_{2\hat{\alpha}}}.$$

Moreover, taking advantage of

$$\text{(5.26)} \qquad \Xi'_G\hat{u}_{\tilde{G}} = \Xi'_G\hat{u}_F,$$

$$\text{(5.27)} \qquad \hat{u}'_{\tilde{G}}\hat{u}_{\tilde{G}} - T\beta_{2\hat{\alpha}} = 2\hat{u}'_{\tilde{G}}\hat{u}_F,$$

$\hat{q}$ and $\hat{l}$ may also be written as

$$\text{(5.28)} \qquad \hat{q} = \frac{1}{T}\begin{pmatrix}\dfrac{\Xi'_G\hat{u}_F}{\beta_{2\hat{\alpha}}} \\[2mm] \dfrac{\hat{u}'_{\tilde{G}}\hat{u}_F}{\beta_{2\hat{\alpha}}^2}\end{pmatrix},$$

$$\text{(5.29)} \qquad \hat{l} = -\frac{1}{2}\log\frac{\beta_{2\hat{\alpha}}}{\hat{\beta}_2} + \frac{1}{T}\frac{\hat{u}'_{\tilde{G}}\hat{u}_F}{\beta_{2\hat{\alpha}}}.$$

Note the resemblances between (5.24)–(5.29) and (3.38)–(3.43).

6. Encompassing tests

Simple sample analogues are considered here, which give consistent estimators of the covariance matrices V_ϕ and V_q, and their Moore-Penrose inverses. This, in turn, yields consistent encompassing tests with correct asymptotic size. The usual asymptotic equivalences are noted, and calculation of the test statistics by means of artificial regressions is discussed. Finally, some remarks are made regarding the asymptotic size of standard non-nested tests.

6.1. Covariance matrix estimation

For the sake of completeness, we first give consistent estimators of the matrices H_F, H_G, $\tilde{H}_G$, $\bar{H}_G$ and B, given in Subsection 4.1. Define

$$\text{(6.1)} \qquad \hat{H}_F = \begin{pmatrix}\dfrac{1}{T}\dfrac{\Xi'_F\Xi_F}{\hat{\alpha}_2} & 0 \\[3mm] 0 & \dfrac{1}{2\hat{\alpha}_2^2}\end{pmatrix},$$

$$\text{(6.2)} \qquad \hat{H}_G = \begin{pmatrix}\dfrac{1}{T}\dfrac{\Xi'_G\Xi_G}{\hat{\beta}_2} & 0 \\[3mm] 0 & \dfrac{1}{2\hat{\beta}_2^2}\end{pmatrix},$$

$$(6.3) \qquad \hat{\tilde{H}}_G = \frac{1}{T} \begin{pmatrix} \dfrac{\Xi_G' \Xi_G}{\beta_{2\hat{\alpha}}} & \dfrac{\Xi_G' \hat{u}_{\tilde{G}}}{\beta_{2\hat{\alpha}}^2} \\[2ex] \dfrac{\hat{u}_{\tilde{G}}' \Xi_G}{\beta_{2\hat{\alpha}}^2} & \dfrac{2 \hat{u}_{\tilde{G}}' \hat{u}_{\tilde{G}} - T \beta_{2\hat{\alpha}}}{2 \beta_{2\hat{\alpha}}^3} \end{pmatrix},$$

$$(6.4) \qquad \hat{\bar{H}}_G = \begin{pmatrix} \dfrac{1}{T} \dfrac{\Xi_G' \Xi_G}{\beta_{2\hat{\alpha}}} & 0 \\[2ex] 0 & \dfrac{1}{2 \beta_{2\hat{\alpha}}^2} \end{pmatrix},$$

$$(6.5) \qquad \hat{B} = \begin{pmatrix} (\Xi_G' \Xi_G)^{-1} \Xi_G' \Xi_F & 0 \\[1ex] \frac{1}{T} 2 \hat{u}_{\tilde{G}}' \Xi_F & 1 \end{pmatrix}.$$

Then, considering (4.7)–(4.10) and (4.16), it follows from Assumptions A1 and A4 that $\hat{H}_F \xrightarrow[\pi_D]{a.s.} H_F$, $\hat{H}_G \xrightarrow[\pi_D]{a.s.} H_G$, $\hat{\tilde{H}}_G \xrightarrow[\pi_D]{a.s.} \tilde{H}_G$, $\hat{\bar{H}}_G \xrightarrow[\pi_D]{a.s.} \bar{H}_G$, and $\hat{B} \xrightarrow[\pi_D]{a.s.} B$.

The estimation of A_ϕ, A_q and A_l is equally straightforward. Putting

$$(6.6) \qquad \hat{A}_\phi = \begin{pmatrix} T (\Xi_G' \Xi_G)^{-1} \\[1ex] -2 \hat{\beta}_1' \end{pmatrix},$$

$$(6.7) \qquad \hat{A}_q = \begin{pmatrix} \beta_{2\hat{\alpha}}^{-1} I_n \\[1ex] -\beta_{2\hat{\alpha}}^{-2} \hat{\beta}_1' \end{pmatrix},$$

$$(6.8) \qquad \hat{A}_l = \frac{T (\Xi_G' \Xi_G)^{-1}}{\hat{\beta}_2} + 2 \frac{\hat{\beta}_1 \hat{\beta}_1'}{\hat{\beta}_2^2},$$

it follows from (4.29), (4.36) and (4.45) and Assumptions A1 and A4 that $\hat{A}_\phi \xrightarrow[\pi_D]{a.s.} A_\phi$, $\hat{A}_q \xrightarrow[\pi_D]{a.s.} A_q$, and $\hat{A}_l \xrightarrow[\pi_D]{a.s.} A_l$.

Recalling that $\Omega = E_P E_D (\tilde{W} U_F^2 W')$, a strongly consistent estimator of Ω under Assumptions A1–A4 is given by

$$(6.9) \qquad \hat{\Omega} = \frac{1}{T} \sum_{t=1}^T \hat{u}_{Ft}^2 \hat{w}_t \hat{w}_t',$$

where

$$(6.10) \qquad \hat{w}_t = x_{Gt} - \Xi_G' \Xi_F (\Xi_F' \Xi_F)^{-1} x_{Ft}.$$

It follows that, under the same set of assumptions,

$$(6.11) \qquad \hat{V}_\phi = \hat{A}_\phi \hat{\Omega} \hat{A}_\phi' \xrightarrow[\pi_D]{a.s.} V_\phi,$$

129

$$(6.12) \qquad \hat{V}_q = \hat{A}_q \hat{\Omega} \hat{A}_q' \xrightarrow[\pi_D]{a.s.} V_q,$$

$$(6.13) \qquad \hat{V}_l = \hat{A}_l \hat{\Omega} \xrightarrow[\pi_D]{a.s.} V_l.$$

Moreover, $\mathrm{rank}(\hat{V}_\phi) = \mathrm{rank}(\hat{V}_q) = r$ with P-probability 1, which ensures that

$$(6.14) \qquad \hat{V}_\phi^+ \xrightarrow[\pi_D]{a.s.} V_\phi^+,$$

$$(6.15) \qquad \hat{V}_q^+ \xrightarrow[\pi_D]{a.s.} V_q^+.$$

As a final point, note that we have not used the assumption $\mathcal{F} \; \mathcal{E}_D^P \; \mathcal{G}$. Rather, we have treated the expressions for A_ϕ, A_q, A_l, Ω, V_ϕ, V_q and V_l as mere definitions and devised consistent estimators for them. For example, $\hat{V}_\phi \xrightarrow[\pi_D]{a.s.} V_\phi$ also if $\mathcal{F} \; \not{\mathcal{E}}_D^P \; \mathcal{G}$, but then V_ϕ is no longer the asymptotic covariance matrix of $\sqrt{T}(\hat{\phi} - \phi)$.

6.2. Encompassing test statistics and critical regions

We consider the following Wald, score and modified likelihood ratio statistics:

$$(6.16) \qquad \tau_\phi = T\hat{\phi}' \hat{V}_\phi^+ \hat{\phi},$$

$$(6.17) \qquad \tau_q = T\hat{q}' \hat{V}_q^+ \hat{q},$$

$$(6.18) \qquad \tau_l = -2T\hat{l}.$$

For any level of significance ϵ between 0 and 1 and any non-zero vector $\lambda \geq 0$, the critical value $c_\epsilon(\lambda)$ of the weighted sum of chi-squares distribution $M(\lambda)$ is defined by

$$(6.19) \qquad \Pr[Z \geq c_\epsilon(\lambda)] = \epsilon,$$

where the random variable Z has distribution $M(\lambda)$. Let $\lambda(V)$ denote the eigenvalues of V, and ι_r an r-vector of ones. The Wald, score and modified likelihood ratio encompassing tests are then defined by the critical regions

$$(6.20) \qquad R_\phi = R_q = [c_\epsilon(\iota_r), +\infty),$$

(6.21)
$$R_l = [c_\epsilon(\lambda(\hat{V}_l)), +\infty),$$

associated with τ_ϕ, τ_q and τ_l. R_ϕ and R_q are simply the $1 - \epsilon$-quantile of the χ_r^2 distribution, while R_l is the $1 - \epsilon$-quantile of a weighted sum of chi-squares distribution.

From the results of Chapter 3, Section 6, it follows that the tests $\langle \tau_\phi, R_\phi \rangle$, $\langle \tau_q, R_q \rangle$ and $\langle \tau_l, R_l \rangle$ are tests of $H_\mathcal{E} : \mathcal{F} \mathcal{E}_D^P \mathcal{G}$ with asymptotic level ϵ and with implicit null hypothesis characterized by $\mathcal{F} \mathcal{E}_D^P \mathcal{G}$. In other words, given Assumptions A1–A4,

(6.22)
$$\mathcal{F} \mathcal{E}_D^P \mathcal{G} \iff \lim_{T \to \infty} \text{Pr}_{\pi_D}[\tau \in R] = \epsilon$$

and

(6.23)
$$\mathcal{F} \mathcal{E}_D^P \mathcal{G} \iff \lim_{T \to \infty} \text{Pr}_{\pi_D}[\tau \in R] = 1,$$

in generic notation. The computation of the Wald and score tests is straightforward. In order to compute the modified likelihood ratio test, one needs the $1 - \epsilon$-quantile of a weighted sum of chi-squares distribution. This may be obtained by simulation or by Imhof's (1961) method, for example.

6.3. Asymptotic equivalences

Under $H_\mathcal{E}$, many asymptotically equivalent tests exist. For example, it follows readily from Theorem 3.4.4 that τ_ϕ and τ_q are asymptotically equivalent, and so are the ensued tests $\langle \tau_\phi, R_\phi \rangle$ and $\langle \tau_q, R_q \rangle$. Further, replacing $\beta_{2\hat{\alpha}}$ by $\hat{\beta}_2$ in $\hat{A}_q$ and thus implicitly in τ_q yields an asymptotically equivalent score test. Similarly, $\hat{\beta}_2$ may be replaced by $\beta_{2\hat{\alpha}}$ in $\hat{A}_l$ and R_l to obtain an modified LR test which is asymptotically equivalent to $\langle \tau_l, R_l \rangle$. The test $\langle \tau_l, R_l \rangle$ itself is not asymptotically equivalent to $\langle \tau_\phi, R_\phi \rangle$ or $\langle \tau_q, R_q \rangle$.

The limiting normal distribution of $\sqrt{T}\hat{\phi}$ is degenerate: V_ϕ is $(n+1) \times (n+1)$ and has rank $r \le n$. As a consequence, asymptotically equivalent encompassing tests may be constructed from quadratic forms in subvectors of $\sqrt{T}\hat{\phi}$ or, more generally, from quadratic forms in $\sqrt{T}\hat{C}_\phi\hat{\phi}$ for a suitably chosen matrix $\hat{C}_\phi$. The same holds, of course, for $\sqrt{T}\hat{q}$. Sufficient conditions for the asymptotic equivalence of the resulting tests are derived from the following lemmas, stated in slightly more general terms than needed here. V^- denotes any generalized inverse of V, i.e. V^- satisfies $VV^-V = V$.

Lemma 6.1. Let V be $n \times n$, symmetric and non-singular, and let C be $p \times n$. Then

(6.24) $$C'(CVC')^- C = C'(CVC')^+ C.$$

Furthermore,

(6.25) $$C'(CVC')^- C = V^{-1}$$

iff C has full column rank.

Proof. By definition,

(6.26) $$CC'(CC')^- CC' = CC'.$$

By the singular-value decomposition, $C = M\Delta^{1/2}N'$ with M and N semi-orthogonal and Δ diagonal and positive definite. Since $CC' = M\Delta M'$ and $N = C'M\Delta^{-1/2}$, pre- and postmultiplication of (6.26) by $N\Delta^{-1/2}M'$ and $M\Delta^{-1/2}N'$, respectively, yields

(6.27) $$C'(CC')^- C = NN' = C'(CC')^+ C.$$

Then $NN' = N'N = I$ iff C has full column rank. The lemma follows if we write $V = V^{1/2}V^{1/2}$ with $V^{1/2}$ symmetric, and replace C by $CV^{1/2}$ in (6.27). □

Lemma 6.2. Let V be $n \times n$, symmetric and non-singular, and let C and A be $p \times q$ and $q \times n$, respectively. Then

(6.27) $$A'C'(CAVA'C')^+ CA = A'(AVA')^+ A$$

iff A and CA have the same rank.

Proof. Since $A'(AA')^+ A$ is idempotent, it may be written as $A'(AA')^+ A = MM'$ with M semi-orthogonal. Next, write $M'A'C'(CAA'C')^+ CAM = N\Delta N'$, with N orthogonal and Δ diagonal, and let $Q = MN$. Then, noting that $A'(AA')^+ AA' = A'$, it is easy to verify that

(6.27) $$A'(AA')^+ A = QQ'$$

132

and

(6.28)
$$A'C'(CAA'C')^+CA = Q\Delta Q'.$$

Since $A'C'(CAA'C')^+CA$ is idempotent so is Δ. Hence

(6.29)
$$A'C'(CAA'C')^+CA = A'(AA')^+A$$

iff Δ is an identity matrix, which occurs iff A and CA have the same rank. Replacing A by $AV^{1/2}$ completes the proof. □

Consider now a quadratic form of the type

(6.30)
$$\tau_{C\phi} = T\hat{\phi}'\hat{C}'_\phi(\hat{C}_\phi\hat{V}_\phi\hat{C}'_\phi)^-\hat{C}_\phi\hat{\phi},$$

for some $s \times (n+1)$ matrix $\hat{C}_\phi$ satisfying

(6.31)
$$\hat{C}_\phi \xrightarrow[\pi_D]{a.s.} C_\phi$$

with C_ϕ non-stochastic, and

(6.32)
$$\operatorname{rank}(C_\phi V_\phi C'_\phi) = r,$$

(6.33)
$$\lim_{T\to\infty} \Pr_{\pi_D} \left[\operatorname{rank}(\hat{C}_\phi\hat{V}_\phi\hat{C}'_\phi) = r \right] = 1.$$

To determine the limit behaviour of $\tau_{C\phi}$ under $H_\mathcal{E}$, observe that, from (4.26)–(4.29),

(6.34)
$$\sqrt{T}\hat{\phi} = A_\phi G M_F z_T + o_p(1),$$

where $z_T = T^{-1/2}\sum_{t=1}^{T} x_{\cup t} u_{Ft}$ is asymptotically normal with zero mean and covariance matrix $V = E_P E_D(X_\cup U_F^2 X'_\cup)$. By the construction of $X_\cup$ and Assumptions A2–A3, V is non-singular. In consequence,

$$\tau_\phi = T\hat{\phi}'\hat{V}_\phi^+\hat{\phi}$$

(6.35)
$$= z'_T M_F G' A'_\phi (A_\phi G M_F V M_F G' A'_\phi)^+ A_\phi G M_F z_T + o_p(1),$$

so that, by Lemmas 6.1 and 6.2,

(6.36)
$$\tau_{C\phi} = \tau_\phi + o_p(1).$$

133

Similar asymptotic equivalences hold regarding the score statistic τ_ϕ, mutatis mutandis. The rank conditions (6.32)–(6.33) are, of course, crucial in arriving at (6.36). If the rank of both $C_\phi V_\phi C_\phi'$ and $\hat{C}_\phi \hat{V}_\phi \hat{C}_\phi'$ would be $r' < r$, then $\tau_{C\phi}$ would be asymptotically distributed as $\chi^2_{r'}$ under $H_\mathcal{E}$, and the test would be inconsistent against some departures from $H_\mathcal{E}$.

In view of the equivalence $\phi = 0 \iff \phi_1 = 0$, it is natural to choose $\hat{C}_\phi$ so as to eliminate $\hat{\beta}_2 - \beta_{2\hat{a}}$ from $\hat{\phi}$. Letting $\hat{C}_\phi = [I_n, 0]$ and using the Moore-Penrose inverse, for example, yields the statistic

$$(6.37) \quad T\hat{\phi}_1'[T(\Xi_G'\Xi_G)^{-1}\hat{\Omega}(\Xi_G'\Xi_G)^{-1}T]^+\hat{\phi}_1$$

$$= \frac{1}{T}\eta'(I - \Pi_F)\Xi_G\hat{\Omega}^+\Xi_G'(I - \Pi_F)\eta$$

$$= \left(\sum_{t=1}^{T}\hat{u}_{Ft}x_{Gt}'\right)\left(\sum_{t=1}^{T}\hat{u}_{Ft}^2\hat{w}_t\hat{w}_t'\right)^+\left(\sum_{t=1}^{T}\hat{x}_{Gt}\hat{u}_{Ft}\right)$$

$$= \left(\sum_{t=1}^{T}\hat{u}_{Ft}\hat{w}_t'\right)\left(\sum_{t=1}^{T}\hat{u}_{Ft}^2\hat{w}_t\hat{w}_t'\right)^+\left(\sum_{t=1}^{T}\hat{w}_t\hat{u}_{Ft}\right).$$

The two latter ways of writing the statistic are reminiscent of the equivalences $\mathcal{F} \, \mathcal{E}_D^P \, \mathcal{G} \iff E_P E_D(X_G U_F) = 0 \iff E_P E_D(W U_F) = 0$ observed earlier. Observe also that applying the same procedure to $\hat{q}$, by taking $\hat{C}_q = [I_n, 0]$, yields exactly the same statistic as in (6.37).

Further, it is a common finding that the test statistic is a quadratic form in $\hat{U}_F$, and that it is numerically identical to T times the uncentered R^2 obtained from an artificial least squares regression, i.e., of 1 on $\hat{u}_{Ft}\hat{w}_{Gt}$, $t = 1, \ldots, T$. See for example the survey by MacKinnon [1992]. Note the rank deficiency of the artificial regression if $r < n$.

Finally, it is easy to see that the statistic in (6.37) is invariant under reformulations of $H_\mathcal{E}$ of the type $H_\mathcal{E} : \mathcal{F} \, \mathcal{E}_D^P \, \mathcal{G}'$, where $\mathcal{G}'$ is a model associated with regressors X_G' such that the joint span of X_F and X_G' equals that of X_F and X_G. This invariance implies in particular the invariance under non-singular linear transformations of X_F and X_G (i.e. $X_F \mapsto MX_F$, $\alpha_1 \mapsto M^{-1}\alpha_1$, $X_G \mapsto NX_G$, $\beta_1 \mapsto N^{-1}\beta_1$). Moreover, choosing r regressors X_G' which span $S_{X_U}^\perp(X_F)$ yields an identical statistic and avoids the rank deficiency if the statistic is computed from an artificial regression.

6.4. Standard non-nested test statistics and critical values

The more traditional approach in the literature on non-nested hypotheses testing consists of deriving the distribution of statistics like $\hat{\phi}$ or $\hat{q}$ under the assumption of correct specification, i.e. $D \in \mathcal{F}$. As a result of this stronger assumption, exact finite sample results are available in the context of normal linear models. In fact it has been shown that the problem of testing non-nested linear regressions with normal errors can be cast in the framework of nested regression models, so one can apply a standard F-test. See for example Deaton [1982] and Dastoor [1983]. The minimal linear nesting model plays the same role there as in the encompassing tests. Here, we shall study the large sample behaviour of standard statistics, shown to have a limiting χ_r^2 distribution if $D \in \mathcal{F}$, under the weaker hypothesis $H_{\mathcal{E}}$.

Recall that, if $D \in \mathcal{F}$, then $\Omega = E_P E_D (W U_F^2 W') = \sigma^2 E_P (W W')$, with σ^2 non-stochastic. In consequence, the standard non-nested test statistics are

$$(6.38) \quad \tau_\phi^* = \left(\frac{1}{T} \sum_{t=1}^{T} \hat{u}_{Ft}^2 \right)^{-1} \left(\sum_{t=1}^{T} \hat{u}_{Ft} \hat{w}_t' \right) \left(\sum_{t=1}^{T} \hat{w}_t \hat{w}_t' \right)^{+} \left(\sum_{t=1}^{T} \hat{w}_t \hat{u}_{Ft} \right),$$

or some asymptotic equivalent thereof. Hence, under $H_{\mathcal{E}}$,

$$(6.39) \quad \tau_\phi^* \xrightarrow[\pi_D]{d} M \left(\lambda \left((E_P E_D U_F^2)^{-1} [E_P (W W')]^{+} E_P E_D (W U_F^2 W') \right) \right),$$

so that the test $\langle \tau_\phi^*, [c_\epsilon(\iota_r), +\infty) \rangle$ has an implicit null hypothesis equal to $H_{\mathcal{E}}$, and asymptotic size differing in general from ϵ, under $H_{\mathcal{E}}$. The larger the covariances are between $\sigma^2 + (\mu - X_F' \alpha_{1D})^2$ and the elements of $W W'$, the larger the asymptotic size of the test based on τ_ϕ^* and standard critical values. The following examples illustrate this.

Misspecification of $\mathcal{F}$, where yet $\mathcal{F} \mathcal{E}_D^P \mathcal{G}$, can come from various sources. By way of example we consider two simple cases, one with conditional variance misspecification and one with conditional mean misspecification.

Let $m = n = 1$, $\mu(X) = X_F$, $\sigma^2(X) = \sigma^2(X_G)$, and assume for convenience that $E_P (X_F X_G) = 0$. Obviously then, $\mathcal{F} \mathcal{E}_D^P \mathcal{G}$, $W = X_G$, and

$$(6.40) \quad \tau_\phi^* \xrightarrow[\pi_D]{d} \kappa \chi_1^2,$$

135

with

(6.41)
$$\kappa = \frac{E_P\left[\sigma^2(X_G)X_G^2\right]}{E_P\left[\sigma^2(X_G)\right]E_P\left[X_G^2\right]}.$$

The sign of the covariance between $\sigma^2(X_G)$ and X_G^2 determines whether or not κ exceeds one. If for example $E_P\left[\sigma^2(X_G)\right] = |X_G|^k$, then

(6.42)
$$\text{sign}(\kappa - 1) = \text{sign}(k).$$

Thus, for positive (resp. negative) k, the true size of the standard test will be larger (resp. smaller) than the nominal size.

Alternatively, let $m = n = 1$ and $E_P(X_F\mu) = E_P(X_G\mu) = 0$. This is the situation where $\mathcal{F}$ nor $\mathcal{G}$ has any explanatory power regarding μ and, where yet $\mu = 0$, represents an extreme case of misspecification of $\mathcal{F}$ where yet $\mathcal{F} \, \mathcal{E}_D^P \, \mathcal{G}$. Let also $\sigma^2(X) = 1$ and $E_P(X_FX_G) = 0$. Similarly then, (6.40) holds with κ now given by

(6.43)
$$\kappa = \frac{E_P\left[(1+\mu^2)X_G^2\right]}{E_P\left[1+\mu^2\right]E_P\left[X_G^2\right]}.$$

The true size of the standard test now exceeds the nominal size if μ^2 and X_G^2 have positive covariance.

7. Conclusion

We have analyzed the encompassing relation on the class of normal linear models, both as a simplifying device in empirical modelling and from an inferential point of view. A basic characterization of encompassing on this class of models is as follows: a linear model $\mathcal{F}$ for Y, given X, encompasses another linear model $\mathcal{G}$ if and only if the orthogonal projection of μ, the true conditional mean of Y, given X, on the subspace of L_2 spanned by the regressors of $\mathcal{F}$ is equal to the orthogonal projection of μ on the subspace spanned by the joint regressors of $\mathcal{F}$ and $\mathcal{G}$. As a result, inference with respect to the encompassing hypothesis can be carried out by testing $\mathcal{F}$ against the comprehensive model induced by $\mathcal{F}$ and $\mathcal{G}$ in a robust sense, that is by using a robust version, in the sense of White [1982], of the covariance matrix of the estimated coefficients. We have considered various Wald and score statistics, which all have the same asymptotic chi-square distribution and are asymptotically equivalent if the encompassing hypothesis holds, and the modified likelihood ratio statistic, which in general has an asymptotic weighted sum of chi-squares distribution and thus is not asymptotically equivalent to the former statistics. The ensued tests are consistent against any departure from the encompassing hypothesis and have correct asymptotic size, irrespective of whether or not the true distribution of Y, given X, belongs to $\mathcal{F}$. Conventional tests in non-nested regression models, like the F test, share the former property with the tests proposed here, i.e., consistency, but not the latter. They have generally incorrect asymptotic size if $\mathcal{F}$ is misspecified.

Conclusion

According to the encompassing principle, a model should be able to account for the results obtained by other rival models. Using a definition of parametric encompassing which has recently been introduced in the econometrics literature, we have analyzed this principle from two different angles. First, we investigated whether encompassing, defined as a relation between models, can put some structure on the class of models related to a given set of empirical phenomena. Given the coexistence in many sciences of alternative theories seeking to explain the same set of phenomena, there is indeed a strong need for ordering principles according to which alternative theories or models can be formally compared. Secondly, for a given pair of competing parametric models $\mathcal{F}$ and $\mathcal{G}$, we analyzed the possibilities of testing the hypothesis that $\mathcal{F}$ encompasses $\mathcal{G}$. This led to the development of a testing methodology which is general in several respects.

Generally speaking, models are intended to mimic relevant aspects of the true process generating the data. Quite fundamentally, from the true process it is always possible to predict the results obtained from a given model $\mathcal{G}$, relative to the uncertainty involved. Hence, the finding that another model $\mathcal{F}$ can mimic this fundamental feature of the true process is evidence in favour of $\mathcal{F}$, whereas the failure of $\mathcal{F}$ to mimic this feature reveals shortcomings of $\mathcal{F}$ as an approximation to the true process. Encompassing formalizes this idea by defining the results from $\mathcal{G}$ to be the maximum likelihood estimator of its parameter. Consequently, $\mathcal{F}$ is said to encompass $\mathcal{G}$ if the prediction from $\mathcal{F}$ of the maximum likelihood estimator associated with $\mathcal{G}$ is not too distant from this estimator, relative to the uncertainty involved. Importantly, $\mathcal{F}$ need not be correctly specified to exhibit this property, although it does so whenever it is correctly specified. The analysis of the properties of the encompassing relation has revealed its non-transitivity and hence its failure to induce a preordering on models. However, in conjunction with the parsimony principle, according to which a submodel is to be preferred to a larger model if it has all the desirable properties of the larger model, the encompassing principle allows a meaningful reduction of the number of empirical models to be considered. We have shown that, if a smaller model $\mathcal{F}$ encompasses a larger model $\mathcal{G}$, then the set of models $\mathcal{P}$ satisfying $\mathcal{F} \subset \mathcal{P} \subset \mathcal{G}$ have identical encompassing properties vis-à-vis all other models, and hence we only need to retain $\mathcal{F}$ from this set. More generally, if $\mathcal{F}$ encompasses $\mathcal{G}$, we only need to consider $\mathcal{G}_{\mathcal{F}}$, the image of $\mathcal{F}$ in $\mathcal{G}$, from the set of models $\mathcal{P}$ satisfying $\mathcal{G}_{\mathcal{F}} \subset \mathcal{P} \subset \mathcal{G}$. The reduction is even more substantial when $\mathcal{F}$ and $\mathcal{G}$ encompass each other. In conclusion, provided that we take the encompassing and parsimony principles as ordering devices, an encompassed model can be reduced to the image of the encompassing model in the encompassed model.

The question of statistical inference with respect to the hypothesis that a given model encompasses another model prompts the development of a statistical theory of robust testing of nested and non-nested models. It requires the distribution theory to be developed without assuming that the true distribution lies within one of the specified models. Apparently, this question has only recently been addressed in the literature. So far, an asymptotic distribution theory concerning Wald and score vectors has been developed in a general dynamic context, under the assumption that the encompassing hypothesis is true. These results led to consistent Wald and score encompassing tests with correct asymptotic size also under misspecification. We addressed the problem of inference in the much less general static context. On the other hand, we obtained distributional results also under fixed deviations from the encompassing hypothesis. Furthermore, we also derived the limit distribution of a modified likelihood ratio statistic both under the encompassing hypothesis and under deviations from it. The Wald and score vectors were shown to be asymptotically normal in general, while the likelihood ratio has an asymptotic weighted sum of chi-squares distribution under the encompassing hypothesis, and an asymptotic normal distribution otherwise. In line with this finding, appropriate quadratic forms in the Wald and score vectors have asymptotic weighted sum of chi-squares distributions under the encompassing hypothesis, and asymptotic normal distributions otherwise. A large class of consistent encompassing tests were deduced from these results. This class comprises many, if not most, parametric tests available in the literature, at least in a static context. Only a subclass, however, was shown to have correct asymptotic size also under misspecification.

We end this conclusion with an outline of a future research agenda.

First, the regularity conditions sustaining the asymptotic distribution theory need to be relaxed. Secondly, the question whether a zero asymptotic variance of the encompassing test statistics implies encompassing has to be resolved. Third, a power analysis under fixed as well as local alternatives remains to be developed. In this respect, it would be interesting to investigate the possibilities of gaining in power by taking quadratic forms in Wald and score vectors and the likelihood ratio simultaneously. This would require the joint limit distribution of these statistics under general conditions. Fourth, in line with the above mentioned research, the approach needs to be extended to the general dynamic case. This extension is particularly relevant to econometrics, where one frequently deals with dependent observations. Fifth, the asymptotic distribution theory has to be extended to test statistics based on simulated pseudo-true values. Sixth, it would be of interest to develop test procedures for the case where one wants to test the hypothesis that a given model simultaneously encompasses a number of other models. Seventh, and probably of more direct interest, the results need to be applied to certain classes of parametric models which are frequently used in practice, in order to provide concise and easily interpretable expressions of the various statistics. Eighth, the encompassing framework opens perspectives for robust tests of hypotheses derived from economic theory by means of misspecified models. In this respect, it would be interesting to apply encompassing tests to the integrability conditions of consumer demand theory by means of presumably misspecified demand models.

Appendix A

Derivation of Tables 1.1 and 1.2

Let f be the density function associated with $F = F_\alpha|_{\alpha=(0,1)}$. A similar notation is used for the other families. Then, from (1.5.1)–(1.5.4),

$$(A.1) \qquad f(y) = \frac{1}{2\sqrt{3}} 1_{[-\sqrt{3},\sqrt{3}]}(y),$$

$$(A.2) \qquad g(y) = \frac{1}{\sqrt{2\pi}} \exp(-y^2/2),$$

$$(A.3) \qquad p(y) = \frac{1}{\sqrt{2}} \exp(-\sqrt{2}|y|),$$

$$(A.4) \qquad q(y) = \frac{\pi \exp(-\pi y/\sqrt{3})}{\sqrt{3}[1 + \exp(-\pi y/\sqrt{3})]^2},$$

and

$$(A.5) \qquad \log f(y) = -\frac{1}{2}\log 12, \qquad (y \in [-\sqrt{3},\sqrt{3}]),$$

$$(A.6) \qquad \log g(y) = -\frac{1}{2}\log(2\pi) - \frac{y^2}{2},$$

$$(A.7) \qquad \log p(y) = -\frac{1}{2}\log 2 - \sqrt{2}|y|,$$

$$(A.8) \qquad \log q(y) = \frac{1}{2}\log \frac{\pi^2}{3} - \frac{\pi}{\sqrt{3}}y - 2\log[1 + \exp(-\pi y/\sqrt{3})].$$

The entropies of F, G, P and Q are given by

$$(A.9) \qquad H(F) = \frac{1}{2}\log 12,$$

143

$(A.10)$
$$H(G) = \frac{1}{2} + \frac{1}{2}\log(2\pi),$$

$(A.11)$
$$H(P) = 1 + \frac{1}{2}\log 2,$$

$(A.12)$
$$H(Q) = 2 - \frac{1}{2}\log\frac{\pi^2}{3},$$

—see Johnson–Kotz [1970b]. The relevant derivatives of the log-densities follow from (A.5)–(A.8) as

$(A.13)$
$$\frac{g'(y)}{g(y)} = -y,$$

$(A.14)$
$$\frac{p'(y)}{p(y)} = -\sqrt{2}\,\mathrm{sign}\,y, \qquad (y \neq 0),$$

$(A.15)$
$$\frac{q'(y)}{q(y)} = -\frac{\pi\left(\exp(\pi y/\sqrt{3}) - 1\right)}{\sqrt{3}\left(\exp(\pi y/\sqrt{3}) + 1\right)}.$$

Now we take up the determination of all pseudo-true values relative to F, G, P and Q, and the implied distances, as far as they are not immediate. This requires solving (4.11) and subsequently applying (4.12). Using (A.1)–(A.8) and (A.13)–(A.15), equation (4.11) takes the following forms for the relevant pairs of distributions and families:

$(A.16)$ $(F, \mathcal{G})$:
$$c_2 - \frac{1}{2\sqrt{3}} \int_{-\sqrt{3}}^{\sqrt{3}} \left(\frac{y}{c_2}\right) y\,dy = 0,$$

$(A.17)$ $(F, \mathcal{P})$:
$$c_2 - \frac{1}{2\sqrt{3}} \int_{-\sqrt{3}}^{\sqrt{3}} (\sqrt{2}\,\mathrm{sign}\,y) y\,dy = 0,$$

$(A.18)$ $(F, \mathcal{Q})$: $c_2 - \dfrac{1}{2\sqrt{3}} \displaystyle\int_{-\sqrt{3}}^{\sqrt{3}} \dfrac{\pi\left[\exp\left(\pi y/(\sqrt{3}c_2)\right) - 1\right]}{\sqrt{3}\left[\exp\left(\pi y/(\sqrt{3}c_2)\right) + 1\right]} y\,dy = 0,$

$(A.19)$ $(G, \mathcal{P})$: $c_2 - \dfrac{1}{\sqrt{2\pi}} \displaystyle\int_{-\infty}^{+\infty} (\sqrt{2}\,\mathrm{sign}\,y) \exp(-y^2/2) y\,dy = 0,$

144

$(A.20)$ $(G, \mathcal{Q})$:
$$c_2 - \frac{1}{\sqrt{2\pi}} \int_{-\infty}^{+\infty} \frac{\pi \left[\exp\left(\pi y/(\sqrt{3}c_2)\right) - 1\right]}{\sqrt{3}\left[\exp\left(\pi y/(\sqrt{3}c_2)\right) + 1\right]} \exp(-y^2/2) y \, dy = 0,$$

$(A.21)$ $(P, \mathcal{G})$: $\qquad c_2 - \frac{1}{\sqrt{2}} \int_{-\infty}^{+\infty} \left(\frac{y}{c_2}\right) \exp(-\sqrt{2}|y|) y \, dy = 0,$

$(A.22)$ $(P, \mathcal{Q})$:
$$c_2 - \frac{1}{\sqrt{2}} \int_{-\infty}^{+\infty} \frac{\pi \left[\exp\left(\pi y/(\sqrt{3}c_2)\right) - 1\right]}{\sqrt{3}\left[\exp\left(\pi y/(\sqrt{3}c_2)\right) + 1\right]} \exp(-\sqrt{2}|y|) y \, dy = 0,$$

$(A.23)$ $(Q, \mathcal{G})$: $c_2 - \frac{\pi}{\sqrt{3}} \int_{-\infty}^{+\infty} \left(\frac{y}{c_2}\right) \frac{\pi \left(\exp(\pi y/\sqrt{3}) - 1\right)}{\sqrt{3}\left(\exp(\pi y/\sqrt{3}) + 1\right)} y \, dy = 0,$

$(A.24)$ $(Q, \mathcal{P})$:
$$c_2 - \frac{\pi}{\sqrt{3}} \int_{-\infty}^{+\infty} (\sqrt{2}\, \text{sign } y) \frac{\pi \left(\exp(\pi y/\sqrt{3}) - 1\right)}{\sqrt{3}\left(\exp(\pi y/\sqrt{3}) + 1\right)} y \, dy = 0.$$

There exist no closed form expressions for the integrals appearing in (A.18), (A.20) and (A.22). We have solved these equations numerically. The other solutions follow from elementary calculus. The results are, in obvious notation,

$(A.25)$ $$c_2(F, \mathcal{G}) = 1,$$

$(A.26)$ $$c_2(F, \mathcal{P}) = \sqrt{\frac{3}{2}} = 1.2247,$$

$(A.27)$ $$c_2(F, \mathcal{Q}) = 1.1024,$$

$(A.28)$ $$c_2(G, \mathcal{P}) = \frac{2}{\sqrt{\pi}} = 1.1284,$$

$(A.29)$ $$c_2(G, \mathcal{Q}) = 1.0371,$$

$(A.30)$ $$c_2(P, \mathcal{G}) = 1,$$

$$(A.31) \qquad c_2(P, \mathcal{Q}) = 0.9446,$$

$$(A.32) \qquad c_2(Q, \mathcal{G}) = 1,$$

$$(A.33) \qquad c_2(Q, \mathcal{P}) = \frac{2\sqrt{6}\log 2}{\pi} = 1.0809.$$

The determination of the RHS of (4.12) requires similar calculations. Using (A.1)–(A.12) and (A.25)–(A.33), we find

$$(A.34) \qquad c_0(F, \mathcal{G}) = \frac{1}{2} - \frac{1}{2}\log\frac{6}{\pi} = 0.1765,$$

$$(A.35) \qquad c_0(F, \mathcal{P}) = 1 - \log 2 = 0.3069,$$

$$(A.36) \qquad c_0(F, \mathcal{Q}) = 0.2217,$$

$$(A.37) \qquad c_0(G, \mathcal{P}) = \frac{1}{2} - \log\frac{\pi}{2} = 0.0484,$$

$$(A.38) \qquad c_0(G, \mathcal{Q}) = 0.0095,$$

$$(A.39) \qquad c_0(P, \mathcal{G}) = \frac{1}{2}\log\pi - \frac{1}{2} = 0.0724,$$

$$(A.40) \qquad c_0(P, \mathcal{Q}) = 0.0219,$$

$$(A.41) \qquad c_0(Q, \mathcal{G}) = \frac{1}{2}\log\frac{2\pi^3}{3} - \frac{3}{2} = 0.0144,$$

$$(A.42) \qquad c_0(Q, \mathcal{P}) = \log\log 16 - 1 = 0.0198.$$

Appendix B

Derivation of formulae (1.5.7)–(1.5.16)

Let f and g be the density functions associated with $F = F_\alpha|_{\alpha_2=1}$ and $G = G_\beta|_{\beta_2=1}$, respectively. From (1.5.5)–(1.5.6), we have

$$(B.1) \qquad f(y) = \frac{1}{\Gamma(\alpha_1)} y^{\alpha_1-1} \exp(-y),$$

$$(B.2) \qquad g(y) = \frac{1}{\sqrt{2\pi\beta_1}y} \exp\left[-\frac{(\log y)^2}{2\beta_1}\right],$$

and

$$(B.3) \qquad \log f(y) = -\log\Gamma(\alpha_1) - (1-\alpha_1)\log y - y,$$

$$(B.4) \qquad \log g(y) = -\frac{1}{2}\log(2\pi) - \frac{1}{2}\log\beta_1 - \log y - \frac{(\log y)^2}{2\beta_1}.$$

We will need the expectations of the log-densities relative to F and G. For this purpose, we use some well-established properties of the gamma distribution which are conveniently expressed in terms of the digamma and trigamma functions. Based on the gamma function

$$(B.5) \qquad \Gamma(a) = \int_0^\infty x^{a-1} e^{-x}\, dx, \qquad a > 0,$$

the digamma and trigamma functions are defined as

$$(B.6) \qquad \psi(a) = \frac{d\log\Gamma(a)}{da}, \qquad a > 0,$$

and

$$(B.7) \qquad \psi'(a) = \frac{d\psi(a)}{da}, \qquad a > 0,$$

147

respectively. See Abramowitz–Stegun [1970, p. 257–60] for mathematical properties of these functions. It follows from properties of the gamma distribution that

$$(B.8) \qquad E_F Y = \alpha_1,$$

$$(B.9) \qquad E_F \log Y = \psi(\alpha_1),$$

$$(B.10) \qquad E_F \left(\log Y - \psi(\alpha_1) \right)^2 = \psi'(\alpha_1),$$

and from properties of the lognormal distribution that

$$(B.11) \qquad E_G Y = \exp(\beta_1/2),$$

$$(B.12) \qquad E_G \log Y = 0,$$

$$(B.13) \qquad E_G (\log Y)^2 = \beta_1.$$

The entropies of F and G follow now from (B.3)–(B.4), (B.8)–(B.9) and (B.11)–(B.12) as

$$(B.14) \qquad H(F) = \log \Gamma(\alpha_1) + (1 - \alpha_1)\psi(\alpha_1) + \alpha_1,$$

$$(B.15) \qquad H(G) = \frac{1}{2} \log \beta_1 + \frac{1}{2} \log(2\pi) + \frac{1}{2}.$$

Since α_2 and β_2 are scale parameters, from (1.2.4) we obtain

$$
\begin{aligned}
I(F, G_\beta) &= -H(F) + \log \beta_2 - E_F \log g(Y/\beta_2) \\
(B.16) \qquad &= -H(F) + \frac{1}{2} \log(2\pi) + \frac{1}{2} \log \beta_1 + \psi(\alpha_1) \\
&\qquad + \frac{\psi'(\alpha_1) + [\psi(\alpha_1) - \log \beta_2]^2}{2\beta_1},
\end{aligned}
$$

$$
\begin{aligned}
I(G, F_\alpha) &= -H(G) + \log \alpha_2 - E_G \log f(Y/\alpha_2) \\
(B.17) \qquad &= -H(G) + \alpha_1 \log \alpha_2 + \log \Gamma(\alpha_1) + \frac{1}{\alpha_2} \exp(\beta_1/2),
\end{aligned}
$$

148

using (B.3)–(B.4) and (B.8)–(B.13). The expressions (B.16)–(B.17) define the minimands involved in the determination of the pseudo-true values relative to F and G. These pseudo-true values are found as

$$(B.18) \qquad \arg \min_{\beta \in R \times R_{++}} I(F, G_\beta) = \left(\psi'(\alpha_1), \exp\left(\psi(\alpha_1) \right) \right),$$

$$(B.19) \qquad \arg \min_{\alpha \in R_{++} \times R_{++}} I(G, F_\alpha) = \left(\kappa(\beta_1), \exp\left(-\psi\left(\kappa(\beta_1) \right) \right) \right),$$

where $\kappa(\beta_1)$ is the functional solution of

$$(B.20) \qquad \log(\kappa) - \psi(\kappa) - \frac{1}{2}\beta_1 = 0.$$

By an appropriate extension of Lemma 1.4.1, we obtain formulae (1.5.7)–(1.5.8) immediately from (B.18)–(B.19). Finally, evaluating the obtained minima, whereby (B.14)–(B.15) are used, the results in (1.5.10)–(1.5.11) follow.

To obtain the limit in (1.5.12), we use the following recurrence formulae

$$(B.21) \qquad \Gamma(a + 1) = a\Gamma(a),$$

$$(B.22) \qquad \psi(a + 1) = \psi(a) + \frac{1}{a},$$

$$(B.23) \qquad \psi'(a + 1) = \psi'(a) - \frac{1}{a^2},$$

and the fact that $\Gamma(1) = 1$ and $\psi(1)$ and $\psi'(1)$ are finite. The limit in (1.5.13) is easily verified with the use of the following asymptotic approximations as $a \to +\infty$:

$$(B.24) \qquad \log \Gamma(a) = (a - \tfrac{1}{2}) \log a - a + \tfrac{1}{2} \log(2\pi) + o(1),$$

$$(B.25) \qquad \psi(a) = \log a - \frac{1}{2a} + o(a^{-1}),$$

$$(B.26) \qquad \psi'(a) = \frac{1}{a} + o(a^{-1}),$$

149

which all derive from Stirling's formula—see Abramowitz–Stegun [1970, p. 257–60]. To verify (1.5.14)–(1.5.15), observe from (1.5.9) that $\kappa(\beta_1) \to +\infty$ (resp. $\kappa(\beta_1) \to 0$) as $\beta_1 \to 0$ (resp. $\beta_1 \to +\infty$), since $\log \kappa - \psi(\kappa)$ is monotonically decreasing in κ. Therefore, the limits in (1.5.14)–(1.5.15) may be restated as limits for $\kappa \to +\infty$ and $\kappa \to 0$, respectively, with β_1 replaced by $2\big(\log \kappa - \psi(\kappa)\big)$. Then, (1.5.14) and (1.5.15) eventually follow upon using (B.24)–(B.25) and (B.21)–(B.23), respectively.

As concerns (1.5.16), the LHS is equal to $\Phi(a)$ since the standardized gamma distribution tends to the standard normal distribution as $\alpha_1 \to +\infty$—see, e.g., Johnson–Kotz [1970a]. As for the RHS, notice that, from (1.5.6)–(1.5.7),

$$\mathrm{Pr}_{\beta_\alpha}\left[\frac{y - \alpha_1\alpha_2}{\sqrt{\alpha_1}\alpha_2} < a\right] = \mathrm{Pr}_{\beta_\alpha}\left[y < \alpha_1\alpha_2 + \sqrt{\alpha_1}\alpha_2 a\right]$$

$$(B.27) \qquad\qquad = \Phi\left[\frac{\log(\alpha_1\alpha_2 + \sqrt{\alpha_1}\alpha_2 a) - \psi(\alpha_1) - \log\alpha_2}{\sqrt{\psi'(\alpha_1)}}\right].$$

Using the approximations (B.25)–(B.26), it can be seen that the RHS of (B.27) tends to $\Phi(a)$ as $\alpha_1 \to \infty$, wherefrom (1.5.16) obtains.

References

Abramowitz, M. and I.A. Stegun (1970). *Handbook of Mathematical Functions*, Dover Publications, New York.

Akaike, H. (1973). "Information Theory and an Extension of the Maximum Likelihood Ratio Principle", in *2nd International Symposium on Information Theory*, ed. by B.N. Petrov and F. Csaki, 267–281, Akademini Kiado, Budapest.

Amemiya, T. (1980). "Selection of Regressors", *International Economic Review*, **21**, 331–354.

Andrews, D.W.K. (1987). "Asymptotic Results for Generalized Wald Tests", *Econometric Theory*, **3**, 348–358.

Aneuryn-Evans, G. and A. Deaton (1980). "Testing Linear versus Logarithmic Regression Models", *Review of Economic Studies*, **47**, 275–291.

Atkinson, A.C. (1970). "A Method for Discriminating between Models", *Journal of the Royal Statistical Society, Series B*, **32**, 323–344.

Barnard, G.A. (1951). "The Theory of Information", *Journal of the Royal Statistical Society, Series B*, **13**, 46–64.

Broze, L. and C. Gouriéroux (1993). "Covariance Estimators and Adjusted Pseudo Maximum Likelihood Method", CORE, Discussion Paper 9313.

Cox, D.R. (1961). "Tests of Separate Families of Hypotheses", in *Proceedings of the Fourth Berkeley Symposium on Mathemathical Statistics and Probability*, Vol. 1, University of California Press, Berkeley, 105–123.

Cox, D.R. (1962). "Further Results on Tests of Separate Families of Hypotheses", *Journal of the Royal Statistical Society, Series B*, **24**, 406–424.

Dastoor, N.K. (1983). "Some Aspects of Testing Non-Nested Hypotheses", *Journal of Econometrics*, **21**, 213–228.

Dastoor, N.K. and M. McAleer (1989). "Some Power Comparisons of Joint and Paired Tests for Nonnested Models under Local Hypotheses", *Econometric Theory*, **5**, 83–94.

Davidson, R. and J.G. MacKinnon (1981). "Several Tests for Model Specification in the Presence of Alternative Hypotheses", *Econometrica*, **49**, 781–793.

Davidson, R. and J.G. MacKinnon (1982). "Some Non-Nested Hypothesis Tests and the Relations among Them", *Review of Economic Studies*, **44**, 551–565.

Deaton, A.S. (1982). "Model Selection Procedures, or, does the Consumption Function Exist?", in *Evaluating the Reliability of Macro-Economic Models*, ed. by G.C. Chow and P. Corsi, John Wiley & Sons, New York, 43–65.

Dhaene, G. (1993). *Encompassing: Formulation, Properties and Testing*, Ph.D. Dissertation, Katholieke Universiteit Leuven.

Engle, R.F., D.F. Hendry and J.-F. Richard (1983). "Exogeneity", *Econometrica*, **51**, 277–304.

Ericsson, N.R. (1986). "Post-Simulation Analysis of Monte Carlo Experiments: Interpreting Pesaran's (1974) Study of Non-Nested Hypothesis Test Statistics", *Review of Economic Studies*, **53**, 691–707.

Fadeev, D.K. (1957). "Zum Begriff der Entropie einer Endlichen Wahrscheinlichkeitsschemas", in *Arbeiten zur Informationstheorie I*, Deutscher Verlag der Wissenschaften, Berlin, 85–90.

Fan, Y. and Q. Li (1995). "Bootstrapping J-Type Tests for Non-Nested Regression Models", *Economics Letters*, **48**, 107–112.

Fisher, G.R. and M. McAleer (1981). "Alternative Procedures and Associated Tests of Significance for Non-Nested Hypotheses", *Journal of Econometrics*, **16**, 103–119.

Godfrey, L.G. and M.H. Pesaran (1983). "Tests of Non-Nested Regression Models. Small Sample Adjustments and Monte Carlo Evidence", *Journal of Econometrics*, **21**, 133–154.

Gouriéroux, C. and A. Monfort (1989). *Statistique et modèles économétriques*, Vol. 1, Economica, Paris.

Gouriéroux, C. and A. Monfort (1992). "Testing, Encompassing and Simulating Dynamic Econometric Models", CREST-INSEE, Discussion Paper 9214.

Gouriéroux, C. and A. Monfort (1994). "Testing Non-Nested Hypotheses", Chapter 44 in *Handbook of Econometrics*, Vol. 4, ed. by R.F. Engle and D.L. McFadden, North-Holland, Amsterdam.

Gouriéroux, C. and A. Monfort (1995). "Testing, Encompassing and Simulating Dynamic Econometric Models", *Econometric Theory*, **11**, 195–228.

Gouriéroux, C., A. Monfort and A. Trognon (1983). "Testing Nested or Non-Nested Hypotheses", *Journal of Econometrics*, **21**, 83–115.

Gouriéroux, C., A. Monfort and A. Trognon (1984). "Pseudo-Maximum Likelihood Methods: Theory", *Econometrica*, **52**, 681–700.

Govaerts, B. (1987). "Application of the Encompassing Principle to Linear Dynamic Models", Ph.D. Dissertation, Université Catholique de Louvain, Louvain-La-Neuve.

Halmos, P.R. (1974). *Measure Theory*, Springer Verlag, New York.

Hendry, D.F., A.R. Pagan and J.D. Sargan (1983). "Dynamic Specification", Chapter 18 in *Handbook of Econometrics*, Vol. 2, ed. by Z. Griliches and M.D. Intriligator, North-Holland, Amsterdam.

Hendry, D.F. and J.-F. Richard (1982). "On the Formulation of Empirical Models in Dynamic Econometrics", *Journal of Econometrics*, **20**, 3–33.

Hendry, D.F. and J.-F. Richard (1983). "The Econometric Analysis of Economic Time Series", *International Statistical Review*, **51**, 111–163.

Hendry, D.F. and J.F. Richard (1990). "Recent Developments in the Theory of Encompassing", in *Contributions to Operations Research and Econometrics, The Twentieth Anniversary of CORE*, ed. by B. Cornet and H. Tulkens, MIT Press, Cambridge.

Huber, P.J. (1967). "The Behaviour of Maximum Likelihood Estimates under Nonstandard Conditions", in *Proceedings of the Fifth Berkeley Symposium on Mathemathical Statistics and Probability*, Vol. 1, ed. by L. LeCam and J. Neyman, University of California Press, Berkeley, 105–123.

Imhof, J.P. (1961). "Computing the Distribution of Quadratic Forms in Normal Variables", *Biometrika*, **48**, 419–426.

Johnson, N.L. and S. Kotz (1970a). *Distributions in Statistics. Continuous Univariate Distributions-1*, John Wiley & Sons, New York.

Johnson, N.L. and S. Kotz (1970b). *Distributions in Statistics. Continuous Univariate Distributions-2*, John Wiley & Sons, New York.

Kent, T. (1982). "Robust Properties of Likelihood Ratio Tests", *Biometrika*, **88**, 19–27.

Kingman, J.F.C. and S.J. Taylor (1977). *Introduction to Measure Theory and Probability*, Cambridge University Press, Cambridge.

Kullback, S. and R.A. Leibler (1951). "On Information and Sufficiency", *Annals of Mathematical Statistics*, **22**, 79–86.

Leamer, E.E. (1978). *Specification Searches: Ad-Hoc Inference with Non-Experimental Data*, John Wiley, New York.

MacKinnon, J.G. (1983). "Model Specification Tests against Non-Nested Alternatives", *Econometric Reviews*, **2**, 85–110.

MacKinnon, J.G. (1992). "Model Specification Tests and Artificial Regressions", *Journal of Economic Literature*, **30**, 102–146.

McAleer, M. (1981). "A Small Sample Test for Non-Nested Regression Models", *Economics Letters*, **7**, 335–338.

McAleer, M. (1983). "Exact Tests of a Model against Nonnested Alternatives", *Biometrika*, **70**, 285–288. in *Specification analysis in the linear model*, ed. by M.L. King and D.E.A. Giles, Routledge and Kegan Paul, London.

McAleer, M. (1987). "Specification Tests for Separate Models: A Survey", in *Specification analysis in the linear model*, ed. by M.L. King and D.E.A. Giles, Routledge and Kegan Paul, London.

McAleer, M. (1995). "The Significance of Testing Empirical Non-Nested Models", *Journal of Econometrics*, **67**, 149–171.

McAleer, M. and M.H. Pesaran (1986). "Statistical Inference in Non-Nested Econometric Models", *Applied Mathematics and Computation*, **20**, 271–311.

Mizon, G.E. (1984). "The Encompassing Approach in Econometrics", Chapter 6 in *Econometrics and Quantitative Economics*, ed. by D.F. Hendry and K.F. Wallis, Basil Blackwell, Oxford.

Mizon, G.E. and J.F. Richard (1986). "The Encompassing Principle and its Applications to Testing Non-Nested Hypotheses", *Econometrica*, **54**, 657–678.

Neyman, J. and E.S. Pearson (1928). "On the Use and Interpretation of Certain Test Criteria for Purposes of Statistical Inference", *Biometrika*, **20A**, 175–240 and 263–294.

Pesaran, M.H. (1974). "On the General Problem of Model Selection", *Review of Economic Studies*, **41**, 153–171.

Pesaran, M.H. (1987). "Global and Partial Non-Nested Hypotheses and Asymptotic Local Power", *Econometric Theory*, **3**, 69–97.

Pesaran, M.H. and A.S. Deaton (1978). "Testing Non-Nested Nonlinear Regression Models", *Econometrica*, **46**, 677–694.

Rao, C.R. (1947). "Large Sample Tests of Statistical Hypotheses Concerning Several Parameters with Applications to Problems of Estimation", *Proceedings of the Cambridge Philosophical Society*, **44**, 50–57.

Rényi, A. (1961). "On Measures of Entropy and Information", in *Proceedings of the Fourth Berkeley Symposium in Mathematical Statistics*, Vol. 1, University of California Press, Berkeley, 547–561.

Richard, J.-F. (1980). "Models with Several Regimes and Changes in Exogeneity", *Review of Economic Studies*, **47**, 1–20.

Sargan, J.D. (1964). "Wages and Prices in the United Kingdom: a Study in Econometric Modelling", in *Econometric Analysis for National Economic Planning*, ed. by P.E. Hart, F. Mills and J.K. Whitaker, Butterworths, London. Reprinted as Chapter 10 in *Econometrics and Quantitative Economics*, 1984, ed. by D.F. Hendry and K.F. Wallis, Basil Blackwell, Oxford.

Sawa, T. (1978). "Information Criteria for Discriminating among Alternative Regression Models", *Econometrica*, **46**, 1273–1292.

Sawyer, K.R. (1983). "Testing Separate Families of Hypotheses: An Information Criterion", *Journal of the Royal Statistical Society, Series B*, **45**, 89–99.

Shannon, C.E. (1948). "A Mathematical Theory of Communication", *Bell System Technical Journal*, **27**, 379–423; 623–656.

Smith, R.J. (1993). "Consistent Tests for the Encompassing Hypothesis", CREST-INSEE, Discussion Paper 9403.

Vuong, Q.H. (1989). "Likelihood Ratio Tests for Model Selection and Non-Nested Hypotheses", *Econometrica*, **57**, 307–333.

Wald, A. (1943). "Tests of Statistical Hypotheses Concerning Several Parameters when the Number of Observations is Large", *Transactions of the American Mathematical Society*, **54**, 426–482.

White, H. (1981). "Consequences and Detection of Misspecified Nonlinear Regression Models", *Journal of the American Statistical Association*, **76**, 419–433.

White, H. (1982). "Maximum Likelihood Estimation of Misspecified Models", *Econometrica*, **50**, 1–26.

White, H. (1994). *Estimation, Inference and Specification Analysis*, Cambridge University Press, Cambridge.

Wiener, N. (1948). *Cybernetics*, John Wiley & Sons, New York.

Wilks, S.S. (1938). "The Large-Sample Distribution of the Likelihood Ratio for Testing Composite Hypothesis", *Annals of Mathematical Statistics*, **9**, 60–62.

Zabel, J.E. (1993). "A Comparison of Nonnested Tests for Misspecified Models using the Method of Approximate Slopes", *Journal of Econometrics*, **57**, 205–232.

Author index

Subject index

Springer
and the
environment

At Springer we firmly believe that an
international science publisher has a
special obligation to the environment,
and our corporate policies consistently
reflect this conviction.
We also expect our business partners –
paper mills, printers, packaging
manufacturers, etc. – to commit
themselves to using materials and
production processes that do not harm
the environment. The paper in this
book is made from low- or no-chlorine
pulp and is acid free, in conformance
with international standards for paper
permanency.

Springer

Lecture Notes in Economics and Mathematical Systems

For information about Vols. 1–264
please contact your bookseller or Springer-Verlag

Vol. 351: A. Lewandowski, V. Volkovich (Eds.), Multiobjective Problems of Mathematical Programming. Proceedings, 1988. VII, 315 pages. 1991.

Vol. 352: O. van Hilten, Optimal Firm Behaviour in the Context of Technological Progress and a Business Cycle. XII, 229 pages. 1991.

Vol. 353: G. Ricci (Ed.), Decision Processes in Economics. Proceedings, 1989. III, 209 pages 1991.

Vol. 354: M. Ivaldi, A Structural Analysis of Expectation Formation. XII, 230 pages. 1991.

Vol. 355: M. Salomon. Deterministic Lotsizing Models for Production Planning. VII, 158 pages. 1991.

Vol. 356: P. Korhonen, A. Lewandowski, J. Wallenius (Eds.), Multiple Criteria Decision Support. Proceedings, 1989. XII, 393 pages. 1991.

Vol. 357: P. Zörnig, Degeneracy Graphs and Simplex Cycling. XV, 194 pages. 1991.

Vol. 358: P. Knottnerus, Linear Models with Correlated Disturbances. VIII, 196 pages. 1991.

Vol. 359: E. de Jong, Exchange Rate Determination and Optimal Economic Policy Under Various Exchange Rate Regimes. VII, 270 pages. 1991.

Vol. 360: P. Stalder, Regime Translations, Spillovers and Buffer Stocks. VI, 193 pages . 1991.

Vol. 361: C. F. Daganzo, Logistics Systems Analysis. X, 321 pages. 1991.

Vol. 362: F. Gehrels, Essays In Macroeconomics of an Open Economy. VII, 183 pages. 1991.

Vol. 363: C. Puppe, Distorted Probabilities and Choice under Risk. VIII, 100 pages . 1991

Vol. 364: B. Horvath, Are Policy Variables Exogenous? XII, 162 pages. 1991.

Vol. 365: G. A. Heuer, U. Leopold-Wildburger. Balanced Silverman Games on General Discrete Sets. V, 140 pages. 1991.

Vol. 366: J. Gruber (Ed.), Econometric Decision Models. Proceedings, 1989. VIII, 636 pages. 1991.

Vol. 367: M. Grauer, D. B. Pressmar (Eds.), Parallel Computing and Mathematical Optimization. Proceedings. V, 208 pages. 1991.

Vol. 368: M. Fedrizzi, J. Kacprzyk, M. Roubens (Eds.), Interactive Fuzzy Optimization. VII, 216 pages. 1991.

Vol. 369: R. Koblo, The Visible Hand. VIII, 131 pages.1991.

Vol. 370: M. J. Beckmann, M. N. Gopalan, R. Subramanian (Eds.), Stochastic Processes and their Applications. Proceedings, 1990. XLI, 292 pages. 1991.

Vol. 371: A. Schmutzler, Flexibility and Adjustment to Information in Sequential Decision Problems. VIII, 198 pages. 1991.

Vol. 372: J. Esteban, The Social Viability of Money. X, 202 pages. 1991.

Vol. 373: A. Billot, Economic Theory of Fuzzy Equilibria. XIII, 164 pages. 1992.

Vol. 374: G. Pflug, U. Dieter (Eds.), Simulation and Optimization. Proceedings, 1990. X, 162 pages. 1992.

Vol. 375: S.-J. Chen, Ch.-L. Hwang, Fuzzy Multiple Attribute Decision Making. XII, 536 pages. 1992.

Vol. 376: K.-H. Jöckel, G. Rothe, W. Sendler (Eds.), Bootstrapping and Related Techniques. Proceedings, 1990. VIII, 247 pages. 1992.

Vol. 377: A. Villar, Operator Theorems with Applications to Distributive Problems and Equilibrium Models. XVI, 160 pages. 1992.

Vol. 378: W. Krabs, J. Zowe (Eds.), Modern Methods of Optimization. Proceedings, 1990. VIII, 348 pages. 1992.

Vol. 379: K. Marti (Ed.), Stochastic Optimization. Proceedings, 1990. VII, 182 pages. 1992.

Vol. 380: J. Odelstad, Invariance and Structural Dependence. XII, 245 pages. 1992.

Vol. 381: C. Giannini, Topics in Structural VAR Econometrics. XI, 131 pages. 1992.

Vol. 382: W. Oettli, D. Pallaschke (Eds.), Advances in Optimization. Proceedings, 1991. X, 527 pages. 1992.

Vol. 383: J. Vartiainen, Capital Accumulation in a Corporatist Economy. VII, 177 pages. 1992.

Vol. 384: A. Martina, Lectures on the Economic Theory of Taxation. XII, 313 pages. 1992.

Vol. 385: J. Gardeazabal, M. Regúlez, The Monetary Model of Exchange Rates and Cointegration. X, 194 pages. 1992.

Vol. 386: M. Desrochers, J.-M. Rousseau (Eds.), Computer-Aided Transit Scheduling. Proceedings, 1990. XIII, 432 pages. 1992.

Vol. 387: W. Gaertner, M. Klemisch-Ahlert, Social Choice and Bargaining Perspectives on Distributive Justice. VIII, 131 pages. 1992.

Vol. 388: D. Bartmann, M. J. Beckmann, Inventory Control. XV, 252 pages. 1992.

Vol. 389: B. Dutta, D. Mookherjee, T. Parthasarathy, T. Raghavan, D. Ray, S. Tijs (Eds.), Game Theory and Economic Applications. Proceedings, 1990. IX, 454 pages. 1992.

Vol. 390: G. Sorger, Minimum Impatience Theorem for Recursive Economic Models. X, 162 pages. 1992.

Vol. 391: C. Keser, Experimental Duopoly Markets with Demand Inertia. X, 150 pages. 1992.

Vol. 392: K. Frauendorfer, Stochastic Two-Stage Programming. VIII, 228 pages. 1992.

Vol. 393: B. Lucke, Price Stabilization on World Agricultural Markets. XI, 274 pages. 1992.

Vol. 394: Y.-J. Lai, C.-L. Hwang, Fuzzy Mathematical Programming. XIII, 301 pages. 1992.

Vol. 395: G. Haag, U. Mueller, K. G. Troitzsch (Eds.), Economic Evolution and Demographic Change. XVI, 409 pages. 1992.

Vol. 396: R. V. V. Vidal (Ed.), Applied Simulated Annealing. VIII, 358 pages. 1992.

Vol. 397: J. Wessels, A. P. Wierzbicki (Eds.), User-Oriented Methodology and Techniques of Decision Analysis and Support. Proceedings, 1991. XII, 295 pages. 1993.

Vol. 398: J.-P. Urbain, Exogeneity in Error Correction Models. XI, 189 pages. 1993.

Vol. 399: F. Gori, L. Geronazzo, M. Galeotti (Eds.), Nonlinear Dynamics in Economics and Social Sciences. Proceedings, 1991. VIII, 367 pages. 1993.